# Teaching Introductory Physics to Visually Impaired Learners

Following global inclusion policies and initiatives, more students with visual impairments are attending regular education in inclusive schools. Universities also increasingly accept students with visual impairments in the most varied courses, including engineering and physics. However, teachers do not always have experience with teaching this specific audience. *Teaching Introductory Physics to Visually Impaired Learners* provides a gateway to understanding the difficulties encountered by this audience and provides ways for teachers to embrace this challenge. The concept of inclusion is discussed in light of the historical evolution of the achievements of people with disabilities and the development of reading and writing in Braille, in addition to other more modern tools, such as cell phone applications. The remaining chapters present methodologies for teaching mechanics, waves, electrodynamics, optics, and modern physics in classes where there are one or more students with visual impairments. The main target audience for this book is elementary and higher education physics teachers and researchers. This book is also of great value to anyone, including teachers from other areas and students with visual impairments, who are looking for teaching materials that facilitate learning for people with visual impairments.

**Key Features:**

- The first book to help readers understand the difficulties of inclusive physics teaching for students with visual impairment.
- Allows the reader to have a broader view of what visual impairment is, bringing together theoretical tools necessary to value and promote inclusive education.

- Encourages the reader to reproduce the physics teaching methodology, with low-cost materials, which facilitates its implementation in the classroom.

**Antônio Carlos Fontes dos Santos** is a Full Professor in the Atomic and Molecular Collisions Laboratory, Department of Nuclear Physics, Institute of Physics, Federal University of Rio de Janeiro. He performs research on ion-atom collisions, multiple ionization, target ionization, and mass spectrometry.

**Marcio Velloso da Silveira** has a postdoctoral degree from the Federal University of Rio de Janeiro. He has a Degree in Physics Teaching from the Federal University of Rio de Janeiro (1997), Specialization in Science and Biology Teaching from Universidade Federal do Rio de Janeiro (2016), Professional Master's Degree in Physics Teaching from UFRJ (2016), and a PhD in Teaching and History of Mathematics and Physics from UFRJ (2021).

# Teaching Introductory Physics to Visually Impaired Learners

Antônio Carlos Fontes dos Santos and
Marcio Velloso da Silveira

CRC Press
Taylor & Francis Group
Boca Raton London New York

CRC Press is an imprint of the
Taylor & Francis Group, an **informa** business

Designed cover image: Shutterstock

First edition published 2026
by CRC Press
2385 NW Executive Center Drive, Suite 320, Boca Raton FL 33431

and by CRC Press
4 Park Square, Milton Park, Abingdon, Oxon, OX14 4RN

*CRC Press is an imprint of Taylor & Francis Group, LLC*



ISBN: 9781032835839 (hbk)
ISBN: 9781032829715 (pbk)
ISBN: 9781003510079 (ebk)

DOI: 10.1201/9781003510079

Typeset in Minion
by codeMantra

# Contents

# Preface

Approximately 16% of the world's population has some form of disability, which corresponds to 1.3 billion people. In developing countries, 90% of children with disabilities do not attend school. In Brazil, for example, among the disabilities, about 45 million people are considered blind, and another 135 million have severe vision limitations. Inclusion laws have been developed over the years, with the Salamanca Statement (1994) being an important milestone in making it a law for governments to promote the socialization of these young people with disabilities, allowing the development of their citizenship.

The laws are implemented very slowly, always falling far short of individual needs, as people with disabilities, being a minority, must learn to live in a world that was not designed for them. However, despite the inertia in which society seems to prefer the blindness of making people with disabilities invisible, technologies have been developed. Small initiatives have proven to be very expensive solutions for those waiting for a world that allows their existence with dignity. Gradually, allowing them to show their values and capabilities, these people have been chosen, with or without government incentive programs, to work in the job market. Thus, with greater autonomy and productivity, these people not only discover a new way of living, despite many setbacks and prejudice from much of society, but also offer substantial contributions to the development of society, helping to cure the social blindness of those who still fail to see people with disabilities as qualified labor, as a workforce, in various fields.

Although laws have been developed and technologies have evolved, there is a huge challenge in meeting the needs of these students in schools and universities. These challenges can be divided into several aspects, but to be clear, without mincing words, the biggest challenge is financial. There is no doubt that education for people with disabilities can be more expensive. This includes special teaching materials, architectural facilities, time, teacher training, and, perhaps the most expensive, empathy.

If you have read this far, you certainly have what is most valuable to meet the needs of students with disabilities, especially visually impaired students.

If you are a physics teacher or in a related field and have just received a visually impaired student, this book can help you not feel so alone on this journey. We will present several tested methods that will allow you to develop low-cost teaching methods and materials to assist in teaching various areas of physics.

We have gathered the most diverse teaching experiences in the field so that you can choose the best way to meet these students' needs. Here is the first tip: get to know your student. Get to know them more deeply, talk to them, and try to understand their journey, the best and worst experiences they had with education, what worked, and what didn't work for them. If possible, talk to the teachers who have already taught them. There is nothing like taking advantage of positive experiences because a classroom is a meeting of constellations, the teacher in their mature universe, and each student, a different constellation in formation.

It is necessary to know your students, and it is even more necessary to know your students with disabilities. But what about time? You will save a lot of time by taking advantage of the experiences that worked and the paths we will provide throughout the next chapters.

The first three chapters will help you understand visual impairment, inclusion, and the general panorama of teaching physics to visually impaired people. If you already have some experience and urgently need to attend to a student in your mechanics classes, go straight to Chapter 4. But return to the first three chapters as soon as you have more time; this can help a lot. Gradually, understanding the needs of visually impaired students, you will be able to develop your own materials and share your experiences for the benefit of the development of physics education for visually impaired people.

## BIBLIOGRAPHY


1. Disability. https://www.who.int/news-room/fact-sheets/detail/disability-and-health. Accessed on January/1/2025.
2. Factsheet on Persons with Disabilities. https://www.un.org/development/desa/disabilities/resources/factsheet-on-persons-with-disabilities.html. Accessed on January/1/2025.
3. 13/12 – Dia do Cego. https://bvsms.saude.gov.br/13-12-dia-do-cego-2/. Accessed on January/1/2025.
4. World Conference on Special Needs Education: Access and Quality, Salamanca, Spain, 7–10 June 1994, United Nations Ministry of Educational, Scientific and Education and Science Cultural Organization Spain, Accessed on https://unesdoc.unesco.org/ark:/48223/pf0000110753.

CHAPTER 1

# Visual impairment

## INTRODUCTION

Once, a visually impaired physics professor expressed his feelings: "...those who have any disability do not have the right to make mistakes, not even the mistakes that those without disabilities make" (free translation) [1, p. 10].

Disabilities can be succinctly classified under two aspects: medical and social. The former is often irreversible, but it is the latter that will determine their future. In other words, disability from a medical standpoint is merely the comparison of the physical and psychological characteristics of an individual against a vast sample of subjects. Within this sample, minorities whose physical and intellectual capacities do not reach a minimum score within a range characterized as "normal" are considered disabled.

From a social perspective, the impact of physical disability can be taken on multiple facets. It can incapacitate an individual for work, leisure activities, and social interaction, or, through the development of medical techniques and technological devices, allow for their complete integration into the world of work, leisure, and social life (inclusion). Below, we will present a more comprehensive discussion on the medical and social impacts of disabilities.

## VISUAL IMPAIRMENT FROM THE MEDICAL PERSPECTIVE

Examining documents from the WHO (World Health Organization), an individual is considered visually impaired when one or more functions of the visual system are affected by a condition of the eye. To determine

DOI: 10.1201/9781003510079-1

whether a person has visual impairment, a visual acuity test is necessary, which involves comparing the distance at which the individual can see with what is considered normal vision, that is, that of a person without any disability [2].

There has been a change in the way patients are examined and diagnosed. Starting in 1972, the diagnosis was made by measuring the best-corrected visual acuity of the patient, that is, using glasses. Visual impairment was diagnosed for individuals who, despite the best correction, had an acuity of less than 6/18 on the Snellen chart in their better eye. Meanwhile, visual impairment was determined by an acuity of less than 3/60 in the better eye. For illustrative purposes, to better understand what this fraction means, consider a visual acuity result of 6/18 in the better eye of the patient. This fraction (6/18) indicates that this patient can only read at a distance of 6 m (standard-sized letters at a predetermined distance), what a person with normal vision, or "healthy eyes," can read at 18 m [2, 3].

However, this classification neglected a large portion of the population who did not have access to refractive correction glasses, leading to a change in the classification of visual impairment in 2010. Patients began to be examined as they presented, that is, whether they were using glasses. In other words, the diagnosis shifted from assessing the best-corrected visual acuity to evaluating the existing visual acuity at the time of the examination. This allowed for the statistical inclusion of a significant portion of the population with visual impairment caused by uncorrected refractive errors. Another change is that previously there was no distinction for varying levels of blindness, for instance, among individuals with acuity lower than 3/60 who either perceive or do not perceive light. From then on, blindness began to be classified into three subcategories, as will be discussed later [3].

The new categories include three subdivisions for individuals considered visually impaired (mild, moderate, and severe), in addition to the diagnosis of blindness. The category "mild impairment vision" includes individuals with visual acuity between 6/12 and 6/18 in their better eye. Those with visual acuity between 6/18 and 6/60 are considered to have moderate vision impairment, and those between 6/60 and 3/60 are classified as having severe vision impairment, both for the better eye. A person is considered visually impaired when their vision in the better eye is worse than 3/60 [3]. It is important to emphasize that even a person classified as visually impaired may, in many cases, be able to perceive light, see shadows, and discern the outlines of people or objects.

## VISUAL IMPAIRMENT FROM THE SOCIAL PERSPECTIVE

The importance and care in analyzing an individual's relationship within the social fabric can be translated by the following quote:

> ...we have the right to be equal when our difference diminishes us; and we have the right to be different when our equality undermines us. Thus, there is a need for an equality that recognizes differences and a difference that does not produce, nurture, or reproduce inequalities. (Free translation)
>
> *[4]*

In the 1970s, in Great Britain, the so-called social model of disability emerged, developed by politicians and intellectuals from UPIAS (Union of the Physically Impaired Against Segregation). This group was formed by people with disabilities inspired by Marxism. The aim of UPIAS was to replace segregating facilities and offer opportunities for individuals with disabilities to participate fully in society, providing them with the chance to engage in productive work that would give them complete control over their lives.

> In our view, it is society which disables physically impaired people. Disability is something imposed on top of our impairments, by the way we are unnecessarily isolated and excluded from full participation in society. Disabled people are therefore an oppressed group in society.
>
> *(UPIAS, 1975, apud [5])*

Beyond Great Britain, many works in Canada and the United States echoed, in the 1990s, the idea of the social model of disability, asserting that the cause of disability lies not within the individual but in the actions that society takes to include this individual so that their citizenship can be fully exercised. In this way, disability only becomes an obstacle for a person in a wheelchair, for instance, if the building they wish to enter lacks ramps or elevators, relying solely on stairs. From this perspective, also developed through the Americans with Disabilities Act of 1990, society is responsible for producing and reproducing disability, as it develops without investing in actions that provide equal opportunities for access to citizenship [6].

The redefinition from the medical model to the social model of disability involves attention to the terminology used to distinguish a personal characteristic from a structural and public phenomenon. In other words, impairment is the physical, organic, or psychological condition that medicine diagnoses through examinations, indicating that an individual does not possess the same abilities as the average or majority of the population. Medicine can, therefore, intervene by diagnosing, reducing, or removing an impairment. On the other hand, disability, from the perspective of the social model, believes that priority should be given to accepting the impairment and removing barriers that prevent the individual from exercising their full citizenship, thereby eliminating the disability [5].

Some researchers believe that the social model of disability, which has been studied by numerous theorists and evidenced through a wide array of publications and research, is grounded in a notion that Foucault would define as "juridic-discursive," that is, underpinned by a legal conception of power, thus becoming predominant in disability studies [7]. According to this author:

> Thus, it would seem that insofar as proponents of the social model claim that disablement is not an inevitable consequence of impairment, they misunderstand the productive constraints of modern power. For it would seem that the category of impairment emerged and in part persists in order to legitimize the disciplinary practices that generated it in the first place.
>
> *[7]*

Davis [8] argues that, although the social model for disability is based on a social and political problem, in the sense that it demands public policies to reduce or eliminate the barriers faced by people with impairments due to societal pressure, the condition of impairment depends on the medical model for its diagnosis. Therein lies a gray area where the occurrence of impairment or disability is not clear. For example, the author asks: "…is Asperger's Syndrome or hysteria an impairment or the creation of the folie à deux of the observing physician and the cooperating patient? Is anorexia or ADD an impairment or a disability?" [8].

Mollow reflects on the characteristic brought by the social model of disability, which differentiates the bodily (impairment) from the social (disability), which can obscure the reality of many illnesses, such as debilitating and terminal diseases or depression, which is a silent and often invisible disease [9].

Activists in the rights of people with disabilities resist individualized treatments seeking cures for bodily impairments, emphatically stating that much of the suffering experienced by these individuals arises from "social exclusion, isolation, and lack of opportunity, along with the often-pernicious side effects of a medical industry bent on aggressive intervention to achieve 'normal' bodies" [10].

Therefore, the social model of disability encompasses a series of dichotomies. First, we can cite the distinction between incapacity, which is an individual aspect, and disability, which is structural and public. Second, there is a distinction between the medical model, which defines disability as an individual deficit, and the social model, which defines disability as a social construct, necessitating the removal of barriers, legislation that reduces discrimination, and promotion of independent living, among other struggles against social oppression. Lastly, the way individuals with disabilities, who form an oppressed group subject to individuals and institutions (including professionals and charities that contribute to this oppression), are distinguished from those without disabilities. In other words, instead of charity, the path to addressing disability is through civil rights. Thus, the social model demonstrates that the challenges faced by individuals with disabilities are a result of social oppression and exclusion, rather than individual deficits, imposing a moral responsibility on society to remove barriers and enable individuals with disabilities to participate [5].

The social model of disability has the power to change perceptions about individuals with disabilities, serving as a psychologically effective model in enhancing the self-esteem of these individuals and in constructing a positive sense of collective identity. Therefore, the problem of disability ceases to be the individual's issue and becomes the problem of the barriers and attitudes that incapacitate them, shifting from the person with a disability to society [5].

The medical model should not be employed as a justification for excluding individuals with disabilities from social practice, but rather to promote individual benefits that can be achieved through specific medical interventions. In other words, medicine should be used to enhance the self-esteem and quality of life for those who feel the need for such interventions. However, this intervention must not occur under social pressure, which tends to convey a message, either explicit or implicit, suggesting that they must conform to a "standard."

Several other social models of disability are emerging, such as the enactive medical model, the ecological-enactive model, and the ecological functional model. Let us understand a bit about what distinguishes these models.

## MODELS FOR DISABILITY DEVELOPED IN RECENT YEARS

We previously discussed the principles regarding the medical and social models of disability. The social model of disability emerges with the aim of clarifying that disability can be minimized or even eliminated through societal initiatives, by means of actions that make the lives of individuals with impairments accessible to various fields of human experience. In other words, society can, through investment and awareness, transform the lives of people with impairment, offering them opportunities to participate in all human activities (leisure, education, work, and social interaction) [11].

Currently, there are several schools of thought and the development of various approaches and frameworks to delineate how disabilities affect individuals. Below, we will look at three of these approaches.

### The enactive medical model

Developed by Michelle Maiese (2021), this model proposes a reformulation of the traditional medical model. Instead of treating disability solely as a biological or physiological issue, Maiese suggests that it should be understood within the context of the interactions between the individual and the environment. The enactive medical model argues that disability arises when the individual's adaptation to the environment is hindered, affecting their ability to engage with and respond to the world functionally [12].

### The ecological-enactive model

Proposed by Toro et al. [13], this model seeks a balance between the medical model and the social model, highlighting the importance of the individual's lived experience. It emphasizes that disability should be seen as an interaction between a person's body and their environment, rather than merely as a pathology. The model suggests that some disabilities may be non-pathological, if the individual can maintain a dynamic adaptation to their surroundings [13].

### The ecological functional model

Created by Robert Chapman [14], this model adopts a relational and pluralistic approach, emphasizing neurodiversity. Chapman argues that disability should be analyzed at three levels: (1) individual mental traits, (2) cognitive styles in the social context, and (3) ecological impact on the group. Instead of focusing on disability as an individual pathology, the

TABLE 1.1 Comparison between the enactive medical, ecological-enactive, and ecological function models

| Criterion | Enactive medical model | Ecological-enactive model | Ecological functional model |
|---|---|---|---|
| **Origin of disability** | Individual adaptation issues | Interaction between body and environment | Relationship between individual and social ecosystem |
| **Main focus** | Maintaining identity and adaptation to surroundings | Lived experience and dynamic adaptation | Neurodiversity and environmental adaptation |
| **Perspective intervention** | Improving individual adaptability | Facilitating adaptation to the environment without pathologization | Modifying the environment to reduce social barriers |
| **Main criticism** | Still maintains a medical and individualistic bias | Difficulty in distinguishing disability from pathology | Risk of relativism by considering disability as ecological variation |

ecological functional model suggests that we should consider the role of the environment in creating or mitigating disability, reinforcing the need to adapt to the environment rather than solely treating the individual [14].

### Comparison of the models

Table 1.1 presents a comparison between the enactive medical, ecological-enactive, and ecological functional models. These frameworks emerge from different philosophical and theoretical traditions and are used to conceptualize health, disability, and agency in embodied and situated terms.

In this way, while the enactive medical model attempts to reformulate the traditional model to include environmental factors, the ecological-enactive model moves further away from the medical perspective by emphasizing the individual's experience. Meanwhile, the functional ecological model proposes a systemic approach, where disability is understood within a social and ecological context.

## OVERCOMING VISUAL IMPAIRMENT

Camargo [15] asserts that visual impairment is considered a social phenomenon, as society has structured itself around a standard of characteristics represented through a "dominant way of being, perceiving, thinking, acting, and living" [15]. However, the person with visual impairment only becomes aware of their condition and the related difficulties imposed by this condition through social interactions, precisely because society is not prepared to incorporate the visually impaired into its culture. Therefore, the barriers faced by the visually impaired are barriers imposed by the sighted society that does not allow the inclusion of those who are different within its social fabric, meaning that it is not the visually impaired individual who should adapt to coexistence with the sighted, as unfortunately continues to occur, but rather society that must provide equal access to all in every possible aspect, so that they can not only enjoy all collective goods, such as scientific and cultural productions, but also allow them to leave their contribution, their legacy, their example, reaffirming that the incapacity does not lie within the individual with a disability, but rather in the limiting perspective of a society that does not believe in their potential.

It is crucial that students with disabilities, facing the difficulties imposed by society, feel represented through access to examples of individuals who, despite their disabilities, have made remarkable contributions to society, achieving significant accomplishments where their impairment did not impose limitations or, conversely, may have enabled access to disparate abilities compared to the sighted. This aligns with what Camargo [15] proposes concerning the understanding of certain physical phenomena as potential advantages. Such dissemination is also beneficial in combating the prejudice that permeates the conscious and unconscious minds of a large part of the population, in order to convince them of the need for investments in inclusion. It is urgent to combat the ideas that persist in society's imagination, where "it is impossible for a human being to know or be happy despite blindness" [16].

The feeling of not belonging can substantially interfere with a student's academic life. Aguilar et al. [17] provide solutions that can assist students who feel unrepresented in the classroom through simple reading and writing interventions, which are activities promoted by teachers for the entire class, aimed at strengthening the sense of belonging of students from underrepresented groups. These interventions should address the concerns of students who feel they do not belong within the school and academic environment, but they should be practiced with the entire group, or all students in the class, so that the intervention does not appear corrective, suggesting that some need more than others [17]. One potential intervention, in this case, could be to present to students, for example, the biographies of notable artists, scientists, and mathematicians who have distinguished themselves or continue to distinguish themselves by contributing to the scientific and cultural advancements of humanity, encompassing different genders and races, and including examples of those with some form of visual impairment. Below, some of these prominent figures will be briefly presented to emphasize that the loss of vision does not equate to an artistic or intellectual incapacity.

One of the seemingly unimaginable examples, which completely contradicts common sense relating to the organic incapacity of a visually impaired person, is the Turkish painter, Ashraf Armagan, who was born visually impaired. Eshref developed his own painting method, where he first outlines on a sheet of paper using a Braille pen, and only then, after the draft is approved, does he transfer the figure to the canvas. The colors, painted with oil paint, are applied with his own fingers, one at a time, allowing approximately three days for each layer to dry before using another color [18]. This innovative method has drawn the interest of many scientists, one of whom provided the following account:

> He is the first person in history to clearly demonstrate that a person born blind can develop normally, without visual contact with the outside world. Nothing like this has ever been recorded, and moreover, there have never been similar cases in the history of the visual arts. (Free translation)
>
> *[18]*

In the fields of mathematics and physics, several names deserve recognition, such as the mathematician Nicholas Saunderson (1682–1739), who became visually impaired in his first year of life due to smallpox. He became fluent in French, Greek, and Latin and eventually became a mathematics professor at the University of Cambridge, holding a position previously held by Isaac Newton and later by Stephen Hawking. His contributions include the creation of a method for performing arithmetic and algebraic calculations, which he called "tangible geometry," using a type of adapted abacus known as Soroban and the Geoboard, both of which are still widely used today [19]. We can also mention the French mathematician Bernard Morin (1931–2018), who lost his sight at the age of six and, during his early childhood, was introduced to a kaleidoscope, which sparked his interest in optical phenomena. Morin developed the ability to imagine an object from the inside out and constructed clay models to represent the inversion of the sphere, which is a highly intriguing problem in a branch of mathematics known as topology and is very difficult to visualize for a sighted person, thanks to his previous experiences [19].

We also have the example of the Swiss mathematician and physicist Leonhard Euler (1707–1783), who became visually impaired in the last 17 years of his life. Nonetheless, he developed half of his 850 works after losing his sight, aided by his prodigious memory and the assistance he received from two of his sons and other members of the Academy of St. Petersburg [19]. The opportunity to learn about these stories of overcoming can be inspiring for students with visual impairments and can help other students and society understand that the loss of vision is not a declaration of incapacity; rather, it should be understood that if society provides opportunities for people with disabilities, they, in turn, can contribute substantially to the progress of society as a whole.

## BIBLIOGRAPHY


1. Camargo, E. P. (2008). *Ensino de Física e deficiência visual: dez anos de investigações no Brasil.* São Paulo: Plêiade/Fapesp.
2. Ben-Eli, H., Banin, E., Levy, J., Glik, M., Afriat, S., Magal, Y., Harari, R., Benyamin, A., Shein, S., & Chowers, I. (2024). Development and evaluation of a new self-administered near visual acuity chart: accuracy and feasibility of usage. *Journal of Clinical Medicine, 13*(7), 2064. https://doi.org/10.3390/jcm13072064.
3. WHO (2019). *World Report on Vision.* World Health Organization, 2011. Available at: https://www.who.int/docs/default-source/documents/publications/world-vision-report-accessible.pdf. Accessed on April/27/2025.
4. Santos, B. S. (2003). *Reconhecer para libertar: os caminhos do cosmopolitanismo multicultural. Introdução: para ampliar o cânone do reconhecimento, da diferença e da igualdade.* Rio de Janeiro: Civilização Brasileira, 56.
5. Shakespeare, T. (2010). The Social Model of Disability. *The Disability Studies Reader.* Ed. Lennard J. Davis. New York: Routledge, 266–273.
6. Davidson, M. (2006). Universal Design: The Work of Disability in an Age of Globalization. *The Disability Studies Reader.* Ed. Lennard J. Davis. New York: Routledge, 117–130.
7. Tremail, S. (2006). On the Government of Disability Foucault, Power, and the Subject of Impairment. *The Disability Studies Reader.* Ed. Lennard J. Davis. New York: Routledge, 185–196.
8. Davis, L. J. (2006). The End of Identity Politics and the Beginning of Dismodernism on Disability as an Unstable Category. *The Disability Studies Reader.* Ed. Lennard J. Davis. New York: Routledge, 231–242.
9. Mollow, A. (2006). When Black Women Start Going on Prozac . . . The Politics of Race, Gender, and Emotional Distress in Meri Nana-Ama Danquah's Willow Weep for Me. *The Disability Studies Reader.* Ed. Lennard J. Davis. New York: Routledge, 283–299.

10. Lewis, B. (2006). A Mad Fight Psychiatry and Disability Activism. *The Disability Studies Reader.* Ed. Lennard J. Davis. New York: Routledge, 339–352.
11. Jurgens, A. (2023). Body social models of disability: examining enactive and ecological approaches. *Frontiers in Psychology, 14,* 1128772. https://doi.org/10.3389/fpsyg.2023.1128772.
12. Maiese, M. (2021). An enactivist reconceptualization of the medical model. *Philosophical Psychology, 34,* 962–988. https://doi.org/10.1080/09515089.2021.1940119.
13. Toro, J., Kiverstein, J., & Rietveld, E. (2020). The ecological-enactive model of disability: why disability does not entail pathological embodiment. *Frontiers in Psychology, 11,* 1162. https://doi.org/10.3389/fpsyg.2020.01162.
14. Chapman, R. (2021). Neurodiversity and the social ecology of mental functions. *Perspectives on Psychological Science, 16,* 1360–1372. https://doi.org/10.1177/1745691620959833.
15. Camargo, E. P. (2008). *Ensino de Física e Deficiência Visual: dez anos de investigações no Brasil,* vol. 1. São Paulo: Editora Plêiade/FAPESP, 10, 17.
16. Bruno, M. M. G. (1999). O significado da deficiência visual na vida cotidiana: análise das representações dos pais-alunos-professores, Dissertação de mestrado, Universidade Católica Dom Bosco, Campo Grande, MS. Available at: https://www.deficienciavisual.pt/txt-Significado_DV_vida_quotidiana-representacoes.htm. Accessed on April/27/2025.
17. Aguilar, L., Walton, G., & Wieman, C. (2014). Psychological insights for improved physics teaching. *Physics Today, 67*(5), 43.
18. Это Удивительно, Слепой художник Эшреф Армаган (2010). https://web.archive.org/web/20100926020225/https://adsence.kiev.ua/2010/09/19/slepojxudozhnik-esref-armagan/. Accessed on April/27/2025.
19. Jackson, A. (2002). The world of blind mathematicians reprinted. *Notices of the American Mathematical Society,* 49(10), 1246–1251. https://www.ams.org/journals/notices/200210/200210FullIssue.pdf. Accessed on April/27/2025.

CHAPTER 2

# What is inclusion?

## INTRODUCTION

As a result of inclusive public policies, the opportunities for students with disabilities to enter and graduate from higher education have increased, which would provide them with greater possibilities for social integration and access to jobs that offer a more dignified financial life [1]. However, there are still several barriers that hinder the integration of these students into higher education, especially the specificities of certain courses, such as the physics course, which accentuates the obstacles for the retention and maintenance of visually impaired (VI) students. The difficulties in accessing and remaining in these programs end up discouraging them, and because of this paradigm, it is observed that students with visual impairments are underrepresented in courses in the fields of Science, Technology, Engineering, and Mathematics (STEM) [2,3].

Students with visual impairments often have difficulties with STEM subjects due to the teaching being predominantly designed for the sighted world. This happens because the way mathematics and other STEM areas are taught is not compatible with teaching for the visually impaired, requiring a series of interventions [3–5]. Some advances can be observed in research in the area, but there is a significant difficulty for such studies to reach the classroom of students with VI which is a reflection of a precarious offer and lack of incentive in the initial and ongoing training of teachers in inclusion area courses [3].

Public inclusion policies have advanced, but the practical effects produced are still slow and involve an effort from the entire society for


DOI: 10.1201/9781003510079-2

inclusion to occur. According to the Salamanca Declaration [6], inclusion is a process aimed at increasing the participation of all students in the learning process, seeking to reduce exclusion in the educational system, which should be achieved through modifications in content, approaches, structures, and strategies, through regular education, which will serve all children of the same age group and through a single educational perspective.

We emphasize that inclusion is related to the education of students with disabilities in the same environment as other students, but that alone is not enough. Between total exclusion and inclusion, there are still two other intermediate forms of access for students with disabilities to education, namely segregation and integration. Figure 2.1 helps us understand the four circumstances that involve education for students with disabilities. According to Villegas [7], "Exclusion occurs when students are directly or indirectly prevented from or denied access to education in any form," and Figure 2.1a exemplifies this situation. Segregation occurs when the education of students with disabilities is provided in a separate environment from students without disabilities and generally, but not necessarily, they are grouped together with students who have the same disability, as can be seen in Figure 2.1b. Integration is the inclusion of students with disabilities in a regular classroom, alongside other students. In this case, it is expected that the students will adjust to the school's standards, not the other way around, as can be seen in Figure 2.1c. Inclusion, as exemplified in Figure 2.1d, goes far beyond integration. For inclusion to occur, the entire school structure needs to be rethought, requiring the development of new teaching and organizational strategies, allowing for the overcoming of barriers and their equal participation in the same environment and age group as the other students [7].

Therefore, the inclusion process involves a lot of planning; it is not enough to simply open the institution's doors and allow access for students with disabilities. However, it only takes a little sensitivity to understand that some situations are potentially prohibitive for Inclusive Education. This is the case for students with comorbidities that do not allow them to remain in the classroom with others, students with severe immunity issues, or those who need to stay in a hospital environment and be assisted by doctors. Or even in global emergencies, such as the COVID-19 pandemic, where social distancing became essential. Except in some extreme situations, efforts should not be spared to ensure the full inclusion of all students, physically in the same school environment and through a school designed and planned to accommodate them. Perhaps this hypothesis of

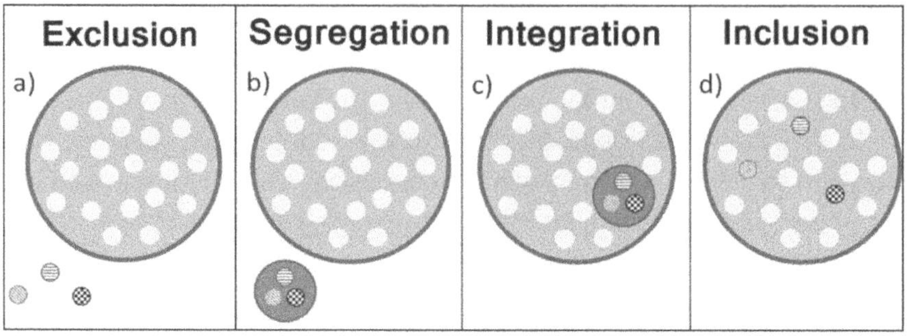


FIGURE 2.1 The white dots inside the larger circles represent students without disabilities placed in a regular classroom, and the textured dots represent students with different types of disabilities. In (a), the textured dots are outside the larger circle, which means that students with disabilities are excluded from the educational system and are not being served. In (b), the textured dots are inside a smaller second circle, which means that students with disabilities are being segregated, that is, they are receiving specialized education, isolated from students without disabilities. In (c), the smaller circle, with the textured dots, is inside the larger circle that contains the white dots, which means that students with disabilities are integrated within the classroom with the other students, but they are still isolated because the school is not adjusted to accommodate them. In (d), the textured dots appear inside the larger circle, mixed with the other white dots, which means that the school has prepared to meet all aspects of their needs, and these students feel included in the class with the other students. This figure is inspired by Villegas [7].

accommodating all students seems somewhat utopian [8], but moving in the opposite direction is to deny what truly makes us human within the fabric of societal living, for in society, no one should be left behind.

## SPECIAL EDUCATION

The special educational approach allows a view of students through the prism of their disabilities, where they are attended to separately from other students, according to their deficits, in an individualized manner by a teacher specially trained to act correctively for their disability. In contrast, Inclusive Education sees the barriers of their disabilities through a human rights approach, understanding that their disability occurs due to external barriers, requiring the teacher to develop a systemic reform in the content, approaches, and educational strategies that enable them to overcome the barriers [9].

Even though today we have the inclusion model as a reference for the education of people with disabilities, it is interesting to briefly analyze the starting point that gradually contributed to the end of sentencing these people to total social and educational exclusion.

The first initiatives to educate people with disabilities date back to 16th-century Spain, with the church and medicine being major influencers in the teaching process for deaf individuals. Monks, who had taken a vow of silence, developed a rudimentary sign language, leading to the creation of the first manual alphabet in history [10].

In the Renaissance period, considered a bridge between the Middle Ages and the Enlightenment, there were significant contributions that led to an increase in scientific knowledge about intellectual disabilities, studies on hydrocephalus, although there also emerged, at this time, claims that linked intellectual disability to witchcraft [11].

The philosophical changes that occurred in France during the Enlightenment period led to the pursuit of a more egalitarian society, based on beliefs about human nature, human rights, human reason, dignity, and self-sufficiency [12]. In England, scholars from the Royal Society of London sought to study language development with the aim of ensuring opportunities for people with disabilities, initially focusing on sensory disabilities, such as the development of oralism as an alternative to sign language, which used lip reading and speech, but which was initially used to serve wealthy deaf individuals, helping them claim their inheritance [12].

The Enlightenment, between the 16th and 18th centuries, was considered an intellectual awakening and featured great philosophers such as Rousseau, Locke, and Condillac who delved into the study of language and sign language, at a time when efforts to educate people with hearing disabilities were increasing, exemplified by the creation of the first oral school for the deaf in Germany in 1755 and the first British Academy for the deaf in 1760 [11].

The development of instructional techniques for serving deaf people encouraged efforts to develop techniques and approaches to serve blind people, leading, in 1784, to the creation of embossed printed books by Valentin Haüy, founder of a school for the blind in Paris [12]. The greatest advancement of the time for blind people was the development of the raised dot system, conceived by a former student of the School for the Blind in Paris, Louis Braille, the Braille system known today, which allowed, through this tactile alphabet, the full inclusion of blind people in French society, spreading its benefits to other countries [12].

The beginning of Special Education is often marked by the work developed with the Wild Boy of Aveyron, whose basic principles, developed by

Jean Marc Gaspard Itard, influenced, in the early 19th century, educators of special schools in Europe and the United States, basing their principles on individualized instructions, carefully crafted task sequences, emphasis on the child's senses, concern for the environment, reward for performance, deepening education as much as possible within individual limits [13].

Special Education emerges to serve children with disabilities who were previously excluded from formal education, but it is not free from criticism, quite the opposite. One of the most obscure points of Special Education is its segregated aspect, as students subjected to this mode of teaching become more susceptible to social isolation and limited opportunities [14]. Inclusive Education researchers accept that the results of the processes are not ideal, but their merit is exemplified through techniques such as direct instruction, mnemonic instruction, self-monitoring, curriculum-based measurement, behavior analysis, functional assessment, and strategic training, which together form a set of techniques that have proven empirically effective in Special Education, with the advantage that many of these techniques can easily be implemented in most traditional classrooms [13].

Special Education depends on very specific technical and human resources, whose specifics are related to disabilities, and, in many cases, a controlled environment so that the senses of students with certain needs are not overloaded [15].

We argue that Special Education was a significant milestone for educational services for people with disabilities in the sense that it sought to reverse a situation of total exclusion. Special Education works in the sense that it is applied with specificity, that is, directed at a specific audience, by teachers trained in this specificity in an environment conducive to this training [13]. Even today, the Special Education system and Inclusive Education coexist, which we will see in a bit more detail below.

## INCLUSIVE EDUCATION

The criticisms of the essentially medical model employed by Special Education have intensified with the advancement of sociological theories that began to interpret disability as a product of environmental barriers [16]. The roots of Inclusive Education trace back to the post-World War II period, when the Universal Declaration of Human Rights (1948) began to recognize education as a fundamental right [17], although it took many decades of questioning from society and academic researchers about the segregation of special schools for any progress to be made.

The fundamental milestone for inclusion was the World Conference on Special Needs Education: Access and Quality (1994), organized by UNESCO, where the principle was established that regular schools must accommodate all children, regardless of their physical, intellectual, and social conditions [18]. As a product of this conference, the Salamanca Statement constituted the cornerstone for educational equity, in line with global policies, determining a profound restructuring of school policies, cultures, and practices.

Other international treaties, such as the Convention on the Rights of Persons with Disabilities (CRPD; United Nations 2006) [19], have contributed to the advancement of Inclusive Education by imposing obligations on the Signatory States to prevent discrimination and ensure human rights concerning education and the inclusion of all children in regular schools [9].

There is an important change in the understanding of disability and the role of the teacher. While in Special Education, students with disabilities were treated separately from other students with corrective measures in an attempt to address their deficits, with the teacher having specialization to specifically deal with the disability, in Inclusive Education the main change is the placement of students with disabilities in regular classes and a reorganization of the curriculum and teaching strategies, with the teacher being the main agent in the implementation of Inclusive Education but requiring support from a specifically organized structure for their work to be developed [9].

Since the Salamanca Declaration, academic interest in Inclusive Education has increased globally, including through the formation of numerous research groups that basically list four schools of thought (systems and structures, critical research, Special Education, and accessibility and participation), whose main topics are as follows: initial teacher training and their attitudes toward Inclusive Education, Inclusive Education in Higher Education, practices and principles of Inclusive Education, and teaching for inclusion and professional development in service of Inclusive Education [20].

Several studies focusing on the theme of inclusion, seeking to understand teachers' perceptions of the process, encounter challenging issues. First, there is the training of teachers in specific skills to address the needs of their students with disabilities, making it safer to cater to these students; second, the perception of the need for an assistant, helping the teacher to fulfill the curriculum that meets all students in the class; there is also a highlight on the perception of teachers working in inclusive classes feeling the need for collaboration from other school professionals and also from other teachers, including those in Special Education, especially regarding the learning specifics of their students in previous years; other points that stood out in the studies were the flexibility of the curricular approach, the size of the class, as they believe that in smaller classes there is more time to pay attention to students with disabilities, the need for specific didactic resources for the practice of Inclusive Education in accordance with the disabilities of their students, and the need for more time, within their workload, for the planning and preparation necessary for the success in the implementation of their lesson in Inclusive Education [9].

We realize that Inclusive Education is a great tool for social inclusion, not only for students with disabilities. The coexistence with the diversity determined by Inclusive Education breaks down taboos and builds, in their place, bridges. Research indicates that students without disabilities socially benefit from interacting in the classroom with students with disabilities [21].

Although varied, the results of most studies on the effects of inclusion for typically developing students demonstrate that the impacts are positive or neutral, and the negative impacts are generally associated with a lack of teacher support and inadequate training, lack of administrative autonomy and support for planning and adapting instruction, cooperation, and low student expectations [21].

The challenges for the implementation of Inclusive Education, as we have seen, are many, but the benefits are evidenced through the observed research. Special Education requires trained professionals, didactic materials, and appropriate environments. Likewise, Inclusive Education does not fulfill its role merely by placing a student with a disability in a regular classroom, expecting the teacher, without any proper training, without the minimum structure, without extended time to prepare their lessons (which must take into account the specific needs of their students with disabilities), without didactic materials that meet the needs, without an assistant, without a myriad of elements that make up Inclusive Education in classrooms with 40 or 50 students. But, unfortunately, this is generally what happens. Inclusion works, but it needs investment that few governments are truly willing to sustain, remaining only in political discourse, its implementation, and the criticism from teachers about the deficiencies in its implementation. And it is up to society to be aware of the foundations of an inclusive society and to hold its representatives in government accountable for providing the resources that can make it effectively viable.

## INCLUSIVE PHYSICS EDUCATION

The teaching of physics is a great challenge due, among other various factors, to its high degree of abstraction and the use of mathematics as a practical tool, even though it is structured in concepts that study and help answer everyday questions such as why objects fall and others, like a helium balloon rising, up to the grand and blue sky [22–24]. While the majority of teachers and researchers are dedicated to finding alternatives that make students interested in learning physics by seeking and reinventing old methodologies with a new approach, few are dedicated to developing methodologies and apparatuses so that blind and visually impaired students can achieve an understanding of such phenomena.

Why teach physics to blind students, since we haven't even found a solution to teach physics to students without disabilities? The answer could be: because there are few people doing this and due to inclusion policies, more and more students with disabilities are entering high school and universities, they need to be served. But that's not all! When we rethink education to cater to students with various disabilities, we end up developing methodologies and tools that can significantly contribute, within Inclusive Education, to the entire class. In other words, not only does the student with a disability benefit from the research, but all the students present in the same classroom do as well [3].

In the following chapters, we will examine approaches to teaching physics to visually impaired students in different areas, with different methodologies and propositions of appropriate teaching materials.

## ASSISTIVE TECHNOLOGIES FOR PEOPLE WITH VISUAL IMPAIRMENTS

Even though it is almost universally agreed today that Inclusive Education surpasses the benefits of Special Education, we must keep in mind the paths that have brought us to the present day, with advances in perceptions of social justice and the development of technologies that allow visually impaired people access to various levels and positions in society. Emerging from a position of total exclusion and vulnerability, in which they were left to their own devices, assisted through charitable institutions, or even exploited in various ways, starting from the 19th century, they began to have a more active participation in society, in various economic, academic, and social activities, transitioning from being assisted to being protagonists of their own destiny and even fighting for the perception of disability to be understood as a lack of means for their direct participation. It means that society itself should strive to provide so that these people can have a voice and a dignified life with protagonism. This struggle is ongoing, and the development of public policies and assistive technologies has opened greater space for the inclusion of people with disabilities, especially those with visual impairments [25].

Valentin Haüy (1745–1822), a linguist and translator fluent in ten languages, was one of the pioneers in developing a tactile method, initially using carved wooden molds where sheets of wet paper were placed and, after drying, painted so they could be read by those with vision. Using his knowledge and techniques, he managed to teach the first blind person to read in three months, using books developed with this technique [26].

Louis Braille (1809–1852), a student at the school for the blind founded by Haüy (Royal Institute for the Blind) in Paris, became blind at the age of 4 and, faced with such a misfortune, believing in the need to educate his son, his father Simon-Rene used upholstery tacks to form letters and manage to teach the letters to the young boy [26]. Later, he managed to enter the Royal Institute for the Blind, being considered an exemplary student. He took advantage of an idea developed by an artillery captain who used a combination of raised dots so they could be read in the dark, and with his ingenuity, he built the six-dot system known today as the Braille writing system, which is widely used and constitutes one of the tools that has most contributed to the educational development of blind people [26].

With the Braille system, students could not only read but also write using the slate and stylus. Later, Braille typewriters emerged. It is important to highlight that despite the significance of the development of this system, we will not elaborate much here to demonstrate its functioning because we believe there is a lot of freely available material that allows for its understanding. In fact, we encourage everyone to dedicate a good amount of time to reading and understanding this system, in order to appreciate it to the point of valuing and helping to spread its importance so that, increasingly, we can connect with the culture of the blind.

Many books have been and still are reproduced in braille, with the disadvantage of taking up several times more space than the original book, thanks to the thickness of each page that uses thicker paper and the embossing of the braille dots. More technological and expensive instruments, such as the Braille line, allow for dozens of Braille characters to be formed by dots that move up and down, giving way to other letters, forming a Braille display. In this way, the entire book, or any text, is being converted into this system of dots while it is being read by the fingers of a blind person. The braille line also allows for writing, similar to a braille typewriter. The equipment is connected to a computer, allowing both the loading of new texts and the writing of texts. Another major advance was the invention of Braille printers, allowing texts to be created on a computer and printed quickly.

For assistance in mathematics, the soroban was created from the abacus, a similar system but in which the beads do not move as easily to prevent a bead from being accidentally moved, causing a calculation error.

It is important to highlight that, just like the sign language used by deaf people, Braille code is not universal; only the alphabetic letters are used worldwide, although, with the same representation, depending on the context, the letter d can also represent a musical note. In other languages, such as Chinese and Hebrew, Braille coding takes on different configurations.

In Brazil, in 1993, a free operating system called DOSVOX, created at the Federal University of Rio de Janeiro, revolutionized the lives of blind students. Today, it has over 100,000 users, and the number of tools keeps increasing. This system features a voice synthesizer that "talks" to the user, allowing the use of a text editor and reader, games, calculators, text translator, and providing access to the internet and social networks, among many other functionalities. The project now includes several other applications that even allow the use of braille dots for creating drawings, tables, and graphs in braille [27].

Many other technologies are also important for the mobility of visually impaired people, such as the cane, guide dog, tactile paving, and Braille signs, which require society to understand and respect their functionality so that visually impaired people can benefit from their use. Others are more recent and technological, such as the sensor cap, electronic cane, robot guide dog, and mobile applications that use GPS to assist in their location.

Other mobile applications, such as By My Eyes, which is also free, allow blind or visually impaired people to request help through the phone's camera to locate objects in the environment, identify medication, among other functionalities. Using AI (Artificial Intelligence), there are several apps that, using the phone's camera, can describe the scene being filmed [28].

Nowadays, these technologies even allow blind people to make bank transactions and shop online, and connected through the Internet of Things, they can turn on and operate devices using voice commands.

Technological evolution is moving toward the development of bionic eyes that could restore vision to blind people. While this reality has not yet arrived, it is important that we keep in mind what we can do today. The way we can adapt the environment so that a person with visual impairment can feel more welcomed.

At this point, we return to the classroom to reflect on how we can make our classes more accessible for people with visual impairments. This reflection and the students' feedback are essential for them to be better served. From the concern with the way of speaking, which should be monitored to avoid "pointing" to something that is not being seen by the visually impaired student, to the development of materials (models, charts, and simulations, among others) that can serve as tactile representations, preferably adapted and visually elaborated to spark the interest of other students.

Another point that deserves attention is mathematical representation, calculation development, and problem-solving. Even though technologies have advanced, the development of algebra, arithmetic, and geometry, which is already a nightmare for most students, often becomes inaccessible for blind students. The problem becomes vertiginously more acute when this student enters a university and faces Calculus. For Calculus, many symbols such as integration and derivation, among others, do not have their development yet mature and well disseminated. This directly impacts the time it takes for this student to complete their education, causing the idea of inclusion to be replaced by a "tentative" attempt at individualized support. Inclusion, in this case, only becomes possible when the visually impaired student begins to master these adaptations, allowing them to develop alongside other students. This means that there is still much to be done for us to achieve success with Inclusive Education at various educational levels [29].

## BIBLIOGRAPHY

1. Sachs, D., & Schreuer, N. (2011). Inclusion of students with disabilities in higher education: performance and participation in student's experiences. *Disability Studies Quarterly, 31*(2), 2011. Retrieved from: https://dsq-sds.org/article/view/1593/156. Accessed on April/27/2025.
2. Toenders, F. G. C., Putter-Smits, L. G. A., Sanders, W. T. M., & Brok, P. (2017). Analysing the physics learning environment of visually impaired students in high schools. *Physics Education, 52,* 1–8.
3. Rule, A. C., Stefanich, G. P., Boody, R., & Peiffer, B. (2011). Impact of adaptive materials on teachers and their students with visual impairments in Secondary Science and Mathematics classes. *International Journal of Science Education, 33*(6), 865–887.

4. Tanti, M. (2007). Teaching mathematics to a blind student - a case study. *Philosophy of Education Journal, 20.* Retrieved from: https://socialsciences.exeter.ac.uk/education/research/centres/stem/publications/pmej/pom e20/Mariella%20Tanti%20Teaching%20Mathematics%20to%20a%20Blind%20Student%20-%20%20%20%20%20%20A%20Case%20Study.pdf. Accessed on April/27/2025.
5. Maindi, A. B. (2018). Challenges faced by students with visual impairments when learning physics in regular secondary schools. *International Journal of Education, 6*(9), 38–50.
6. UNESCO (1994). Salamanca Statement and Framework for Action on Special Needs Education. Paris: United Nations Educational, Scientific and Cultural Organization.
7. Villegas, T. (2017). Inclusion, exclusion, segregation and integration: how are they different? The Official Blog of Maryland Coalition for Inclusive Education. Retrieved from: https://mcie.org/think-inclusive/inclusion-exclusion-segregation-and-integration-how-are-they-different/. Accessedon April/27/2025.
8. Young, K., Mcnamara, P. M., & Coughlan, B. (2017). Authentic inclusion-utopian thinking? – Irish post-primary teachers' perspectives of inclusive education. *Teaching and Teacher Education, 68*, 1–11.
9. Chow, W. S. E., Bruin, K. De, Sharma, U. (2024). A scoping review of perceived support needs of teachers for implementing inclusive education. *International Journal of Inclusive Education, 28*(13), 3321–3340. https://doi.org/10.1080/13603116.2023.2244956.
10. Konkel, E. N., & Miguel, Maria Elisabeth Blanck (2019). INICIATIVAS DE ATENDIMENTO EDUCACIONAL A PESSOA COM DEFICIÊNCIA EM SANTA CATARINA: PERCURSO HISTÓRICO. Notandum.
11. Mutua, K., Siders, J., & Bakken, J. P. (2011). Chapter 5: History of Intellectual Disabilities. *History of Special Education*. Ed. A. F. Rotatori, F. E. Obiakor, & J. P. Bakken. Bingley: Emerald Group Publishing Limited, 89–117. https://doi.org/10.1108/S0270-4013(2011)0000021004.
12. Salend, S. J., & Garrick Duhaney, L. M. (2011). Chapter 1: Historical and Philosophical Changes in the Education of Students with Exceptionalities. *History of Special Education*. Ed. A. F. Rotatori, F. E. Obiakor, & J. P. Bakken. Bingley: Emerald Group Publishing Limited, 1–20. https://doi.org/10.1108/S0270-4013(2011)0000021004.
13. Cook, Bryan & Schirmer, Barbara. (2003). What is special about special education?: overview and analysis. *Journal of Special Education, 37*, 200–205. https://doi.org/10.1177/00224669030370031001.
14. Strogilos, V., Lacey, P., Xanthacou, Y., & Kaila, M. (2011). Collaboration and integration of services in Greek special schools: two different models of delivering school services. *International Journal of Inclusive Education, 15*(8), 797–818.
15. Mu, G. M. et al. (2017). Building resilience of students with disabilities in China. *Teaching and Teacher Education, 67*, 125–134. https://doi.org/10.1016/j.tate.2017.06.004.

16. França, T. H. (2013). Modelo Social da Deficiência: uma ferramenta sociológica para a emancipação social. *Lutas Sociais, 17*(31), 59–73.
17. United Nations. (2025). Universal Declaration of Human Rights. Retrieved from: https://www.un.org/en/about-us/universal-declaration-of-human-rights. Accessed on Jan/1/2025.
18. World Conference on Special Needs Education: Access and Quality, Salamanca, Spain, 7–10 June 1994, United Nations Ministry of Educational, Scientific and Education and Science Cultural Organization Spain, Access in: https://unesdoc.unesco.org/ark:/48223/pf0000110753.
19. United Nations. (2006). Convention on the Rights of Persons with Disabilities (CRPD). https://www.un.org/disabilities/documents/convention/convoptprot-e.pdf. Accessed on April/27/2025.
20. Hernández-Torrano, D., Somerton, M., & Helmer, J. (2020). Mapping research on inclusive education since Salamanca statement: a bibliometric review of the literature over 25 years. *International Journal of Inclusive Education, 26*(9), 1–20. https://doi.org/10.1080/13603116.2020.1747555.
21. Kart, A., & Kart, M. (2021). Academic and social effects of inclusion on students without disabilities: a review of the literature. *Education Sciences,* 11(1), 16. https://doi.org/10.3390/educsci11010016.
22. Moreira, M. A. (2021). Desafios no ensino da física. *Revista Brasileira de Ensino de Física, 43*(suppl 1), e20200451. https://doi.org/10.1590/1806-9126-rbef-2020-0451.
23. Volkmann, M. J., Abell, S. K., & Zgagacz, M. (2005). The challenges of teaching physics to preservice elementary teachers: orientations of the professor, teaching assistant, and students. *Science Education, 89*(5), 847–869. https://doi.org/10.1002/sce.20077.
24. Regra, A. C., Stefanich, G. P., Boody, R. M., & Peiffer, B. (2010). Impacto dos Materiais Adaptativos em Professores e seus Alunos com Deficiência Visual nas Aulas de Ciências e Matemática do Ensino Médio. *Revista Internacional de Educação Científica, 33*(6), 865–887. https://doi.org/10.1080/09500693.2010.506619.
25. Haüy V. (1786). *An Essay on the Education of the Blind.* An explication of the different means, confirmed by successful experiments, to render them capable of reading by the assistance of touch, and of printing books, in which they may obtain the knowledge of languages, of history, of geography, of music, etc. Paris: French Academy of Sciences. Historical Summary. https://play.google.com/books/reader?id=ZIdDAAAAIAAJ&printsec=frontcover&output=reader&hl=en &pg=GBS.PP1. Accessed on April/24/2025.
26. Morse, A. R. (2017). Valentin Haüy and Louis Braille: Enabling Education for the Blind. *Foundations of Ophthalmology.* Ed. M. Marmor, & D. Albert. Cham: Springer. https://doi.org/10.1007/978-3-319-59641-9_5.
27. Ribeiro, A., & Marino, M. (2025). SINTETIZADORES DE VOZ: A TECNOLOGIA E A ACESSIBILIDADE DO DISCENTE. REIN - *REVISTA EDUCAÇÃO INCLUSIVA, 10*(2), 39–52. https://revista.uepb.edu.br/REIN/article/view/3496.

28. Ribeiro, A., & Marino, M. (2024). CULTURA DIGITAL: INOVAÇÃO DOS RECURSOS DE TECNOLOGIA ASSISTIVA VOLTADOS PARA A EDUCAÇÃO DE PESSOAS COM DEFICIÊNCIA VISUAL. *Revista Eletrônica Do Instituto de Humanidades*, 32(58), 143–156. https://granrio.emnuvens.com.br/reihm/article/view/9052.
29. Silva, M. A., & Coelho, C. C. (2022). Ensino de Cálculo a Alunos com Deficiência Visual: a Importância do Professor e do Uso de Recursos de Tecnologia Assistiva. *Perspectivas Da Educação Matemática, 15*(38), 1–21. https://doi.org/10.46312/pem.v15i38.14551.

CHAPTER 3

# Current advances in physics education for students with visual impairments

## INTRODUCTION

Individuals with visual impairments have long been marginalized in classroom and often perceived as a burden. It is fundamental for physics educators to prioritize the unique knowledge, skills, and experiences of disabled individuals and to push for systemic changes that ensure their full inclusion in physics classrooms.

Teaching visually impaired students science, technology, engineering, and mathematics (STEM) is essential for fostering their ability to think logically, explain material concepts through various representations, classify phenomena as examples or non-examples of abstract ideas, translate concepts into mathematical language to solve complex problems, and apply these concepts in everyday life. A strong conceptual understanding is also critical for higher-order thinking skills, such as critical and creative thinking.

In addition to learner-related factors, the use of instructional models that do not align with the specific learning needs of students can


DOI: 10.1201/9781003510079-3

exacerbate these challenges. Therefore, systematic research is needed to review journal articles that examine the effectiveness of teaching strategies for enhancing conceptual understanding in visually impaired students. This approach aims to identify effective methods for improving their grasp of physics concepts and to explore how these strategies are applied in education through a systematic literature review.

Students with visual impairments face challenges in accessing information through their sense of sight. Learning physics must not only be accessible but also prepare students for higher education, employment, and independent living. It should also help students, including those with visual impairments, develop critical thinking skills and problem-solving abilities essential for life.

Students with visual impairments exhibit the same range of cognitive abilities as their sighted peers, and with appropriate accommodations, they can grasp advanced science concepts just as effectively. Several studies have examined how science education can be adapted for students with visual impairment [1–13]. Visual impairments can negatively impact academic performance, with students with visual impairment often achieving lower academic results.

One must underline the need for tailored approaches in teaching physics to students with disabilities. However, a comprehensive literature review is necessary to synthesize previous research findings and guide future studies in science education for students with visual impairments. Such a review would provide valuable insights for improving teaching practices and outcomes in this area.

## PREVIOUS LITERATURE REVIEWS

Castro-Castillo and Tuay-Sigua [1] carried out a review of 27 articles on the state of the art on the teaching of physics in students with visual impairment, with an observation window from 2010 to 2022. This chapter presents an update to their review. Castro-Castillo and Tuay-Sigua [1] highlight significant progress in developing, adapting, and implementing specialized materials and software that enable individuals with visual impairment to access information through haptic and auditory channels. This advancement emphasizes the importance of exploring diverse teaching methods to promote inclusive education, fostering meaningful interaction with these communities and enhancing their understanding of physical world phenomena.

In the context of physics education, the opportunity arises to review, analyze, and propose strategies that allow students to engage with experiences aligned with their sensory abilities. Such approaches enable the active participation of all students in lessons and support the construction of scientific knowledge through cognitive, emotional, and sociocultural dimensions.

Castro-Castillo and Tuay-Sigua [1] grouped their findings into the following categories derived from the analysis: (i) development and adaptation of learning environments and instructional materials (17 studies), (ii) exploration of communication dynamics and environmental interaction in physics classes (2 studies), (iii) teacher training focused on the inclusion of individuals with visual functional diversity (3 studies), (iv) documentary reviews aimed at establishing the state of the art and analyzing existing public policies (2 studies), and (v) representations of blind students regarding specific physics-related topics or their experiences with inclusion in physics classes (3 studies).

Their findings indicate that most beneficiaries of these investigations are young people in secondary education or, in some cases, early higher education. The authors suggest extending these efforts to primary education would be valuable. Early interventions could help understand how children begin to engage with scientific knowledge during their formative years, contributing to the creation of inclusive educational environments that celebrate diversity.

According to Castro-Castillo and Tuay-Sigua [1], most of the research in teaching physics to the blind is concentrated in Brazil, likely due to several factors. The country has made significant advancements in public policies and regulations for special education, as well as initiatives promoting the inclusion of individuals with disabilities. Additionally, numerous school programs have been developed to address the specific needs of students.

Brazil also hosts several active research groups, including one led by Dr. Eder Pires de Camargo [2–6], a teacher with visual functional diversity, who has focused on developing didactic and communicative approaches in the field. Similarly, other authors have contributed by designing tactile resources to support the inclusion of students with visual impairment in physics classes. The authors have explored various strategies, incorporating art, narrative, and language, to help these communities engage with and study phenomena within the discipline.

*Physics topics addressed:* The most frequently addressed physics topics are those related to optics, thermodynamics, conception of space, movement, sound, force, work and power, oscillations, wave phenomena and notions of modern physics and electricity, in which different strategies have been implemented with visually impaired students to link them to the teaching and learning processes of the discipline.

Velloso et al. [7, 8] highlight that engaging visually impaired students in experimental physics is feasible. Activities such as using multimeters, assembling electrical circuits, taking measurements, and even generating graphs can be effectively conducted, provided the teaching strategies are designed with confidence in the students' abilities. It is worth noting that these activities were carried out by a blind student participating in a university-level education program. In a like manner, Lisboa et al. [9] demonstrate that students with visual functional diversity can successfully take measurements and achieve results within acceptable error margins. In their study, this was accomplished using an adapted distance measurement system incorporating a position sensor connected to an Arduino platform.

Ediyanto and Kawai [10] performed an in-depth literature review by analyzing a total of 17 journal articles from the year of 2009–2018 written in English on the topic of science learning for students with visual impairment. The authors concluded that students with visual impairments can excel in science when provided with appropriate support tools. By implementing effective teaching models and ensuring access to adequate resources—such as orientation and mobility aids, tactile and kinesthetic learning tools, auditory accommodations, and assistive technologies—teachers can create an environment where students with VI can thrive in learning science.

Ediyanto and Kawai [10] suggested that assistive technology can significantly enhance students' efficiency, largely depending on the teacher's knowledge and proficiency in using it. Additionally, classroom facilities should be adapted to meet the needs of students with visual impairments. This aligns with the importance of teachers valuing and effectively utilizing assistive technologies for students with visual impairments. Among the most effective tools are Braille embossers and computers. For tactile and kinesthetic learning, students with visual impairments can benefit from direct interaction with objects, such as visiting museums and creating 3D models. Hands-on interaction with objects or models provides students with valuable information and enhances their understanding of the subject matter.

Chini and Scanlon [11] performed a literature review on teaching physics with disabled learners. They analyzed 66 sources. The sources were reviewed and categorized based on their intended audience (education practitioners or STEM education researchers) and their focus (laboratory practices, general education practices, technology, conceptual understanding, or universal design for learning). Identifying relevant literature proved challenging, as few articles referenced related existing sources, indicating that the existing body of research is not well interconnected. The most frequently addressed topic in the reviewed sources was how to adapt existing laboratory equipment to include disabled students in physics labs. These sources explored various strategies to ensure accessibility, categorized into three main areas: (a) modifications to existing equipment, (b) accessible laboratory tools and assistive technologies, and (c) methods and tools for making specific experiments accessible.

The modification to existing equipment includes sources spanning over five decades that detail ways to make physics laboratory equipment accessible to students with visual impairments. The recommendations gathered from these sources for supporting students with visual impairments include the following [11]:

- Using raised lines to provide access to diagrams and graphs.
- Creating tactile representations of key features of laboratory equipment (e.g., tactile metersticks or micrometers which allow users to feel measurement readouts.
- Using magnification of measurement readouts, e.g., large print.
- Using audification (converting output to sound) of measurement readouts, e.g., talking clocks or calculators.
- Adding Braille writing to equipment instructions and readouts as well as demonstrations.

In the second category—accessible laboratory tools and assistive technologies—sources discuss the development of specially designed accessible laboratory tools to help disabled students access experiments, equipment, and/or data.

Pacala [12] conducted a systematic review on the role of technology in teaching physics to students with disabilities. The study aimed to identify the classroom technologies utilized, assess their effectiveness, and determine the constructs these technologies addressed. The review analyzed 13 studies from various journal repositories, with visual impairment and intellectual disabilities being the most studied conditions.

The research predominantly focused on learner achievement, followed by engagement and participation. Assistive technologies, such as web-based learning platforms, JAWS screen-reader software, SEED, tablets, and smartphones, emerged as the primary tools used to support learning. Other technologies included virtual and augmented reality (e.g., Magic Leap, Minecraft) and interactive tools like LEGO Education EV3 Mindstorms. All 13 studies demonstrated the positive impact of these technologies, enhancing student engagement, social inclusion, and enthusiasm for science. AR technology was particularly effective in improving performance and knowledge retention, showing promise as an engaging and accessible tool in inclusive physics education.

The review emphasized a growing trend toward integrating technology in inclusive education to address the unique challenges faced by students with disabilities. The findings highlight the importance of personalized tools in fostering equitable access and interactive learning. Future research should explore the broader implementation and long-term effects of such interventions across diverse educational settings and disability types. This includes developing and accessing innovative resources like 3D virtual labs, mobile technologies, AR tools, and auditory aids.

Research gaps identified include the need for larger and more diverse participant groups to enhance the reliability and applicability of findings. Common limitations included small sample sizes, resource and technological constraints, adaptation challenges, and practical issues. Collaboration across multiple schools and regions could help address these gaps, ensuring broader representation of students with various disabilities.

The review also pointed out the need for targeted investments in accessible technologies and infrastructure, particularly in under-resourced areas. Additionally, future studies should consider participants' age, health status, and the specific physics skills being taught. Expanding research to include students from diverse backgrounds and exploring interdisciplinary connections in physics education are promising directions.

Teacher training programs tailored to the needs of students with disabilities, pilot testing of physics interventions, and longitudinal studies on sustained engagement and knowledge retention are also essential. Scaling up physics education interventions to ensure equitable access for all learners remains a critical priority. By addressing these recommendations, the inclusiveness and quality of physics education for students with disabilities can be significantly enhanced.

Scholars repeatedly report positive outcomes when instructional strategies and techniques are tailored to support the learning needs of blind students. Studies generally fall into two categories: those that create new materials and those that adapt existing ones.

Blind students often acquire pre-instructional knowledge through hands-on experiences, such as laboratory experiments, and resources like Braille books, audio recordings, internet tools equipped with screen readers, and guidance from peers, educators, and experts. However, these resources are insufficient to fully understand how their pre-instructional knowledge differs from that of sighted students.

Research on science education for blind students frequently emphasizes inclusive learning environments. While many educators advocate for inclusive classrooms where students with disabilities learn alongside their peers, some teachers feel unprepared to teach visually impaired students due to limited training and experience. Nevertheless, studies demonstrate the benefits of collaborative learning in inclusive classrooms for visually impaired students. Interactions with sighted peers help blind students develop physical concepts and enhance their understanding of science. However, schools specifically for blind or partially sighted students often focus more on physical accommodation than on cognitive needs.

One significant challenge in inclusive education is comparing the academic achievements of blind and sighted students. Studies show that with adequate time and effective teaching methods, blind students can achieve similar academic success as their sighted peers. Flexible approaches that allow blind students to participate in authentic scientific procedures without time constraints have proven to be particularly effective.

França and Siqueira [14] aimed to identify teaching materials already developed for covering various physics topics for visually impaired students and to analyze trends in these works in Brazil. Their investigation included a literature review of studies focused on producing physics teaching materials for the visually impaired. While they identified a notable number of materials (48), there remains a significant need to create additional resources, particularly for topics like Thermal Physics, Fluid Mechanics, and Modern and Contemporary Physics, which are central to basic education.

The development of physics teaching materials specifically designed for visually impaired students opens numerous opportunities for further work in this area, enabling the study of phenomena through tactile exploration. These materials are designed to utilize the students' sense of touch and ensure safety, fostering engagement and understanding without physical risk. Interestingly, these activities not only benefit visually impaired students but also support sighted students in grasping complex concepts.

The authors conclude that more research is necessary, particularly in regions with limited production of such materials. New research groups and higher education institutions should contribute to advancing knowledge on teaching and learning processes for students with visual impairments. Although many activities have been implemented and evaluated by blind or low-vision students, there is a need for deeper reflection and analysis by researchers. The teaching and learning process must be understood through robust theoretical and methodological frameworks to provide well-founded insights.

Currently, most efforts focus on disseminating activities, which, while valuable, do not comprehensively address the underlying processes of teaching and learning for visually impaired students. A stronger theoretical basis is essential to analyze the development and effectiveness of these activities.

The topic is highly significant for physics education and underscores the importance of creating innovative teaching activities not only in physics but also across other disciplines and for a variety of disabilities or special educational needs. Future research should emphasize implementing and analyzing these activities in classroom settings to enhance the educational experiences and outcomes of students with visual impairments and other special needs.

Oliveira et al. [15] conducted a bibliographic survey of Brazilian journal databases (from 2018 to 2022), identifying studies focused on producing accessible and low-cost materials with significant contributions to inclusive methodologies in physics education.

The analysis of published works revealed that most materials focused on facilitating the understanding of topics such as optics, electromagnetism, waves, and astronomy. However, a notable gap exists in the production of materials for students with other types of disabilities, such as intellectual or hearing impairments, as most studies center on students with visual impairments. This focus is likely due to the inherently visual nature of physics, which relies heavily on figures, tables, and graphs.

The use of teaching materials and resources that enhance learning is critical. Beyond fostering inclusion, these resources also stimulate curiosity and encourage active participation among students with disabilities. For physics teachers, it is crucial to adopt a reflective approach to both content and teaching methodology, remaining attentive to the unique challenges faced by students with disabilities. By tailoring their methods to the specific needs and realities of their students, teachers can create a more inclusive and supportive learning environment.

De Souza Mól et al. [16] address the inclusion of the target audience of special education in regular education, focusing on science teaching. The authors highlight the importance of Inclusive Education as a paradigm that aims to democratize education and ensure equal access for all students, including those with disabilities, global developmental disorders, and giftedness. They present a relevant conceptual review, citing authors who support the necessity of inclusion and discussing how this topic has been approached in scientific research since the early 2000s.

Their study employed a systematic literature review to map the main publications on inclusion in science education. The authors analyzed 59 articles published in 21 top-tier journals, and all editions of the Brazilian Meeting on Research in Science Education (ENPEC) from 1997 to 2019 were reviewed.

The authors' main findings include growth in research on inclusion in science education over the years, particularly after 2000. The authors found a predominant focus on visual and auditory impairments, with less attention given to other special needs, such as global developmental disorders and giftedness. The authors also pointed out a limited emphasis on teacher training, indicating a gap in teacher preparation for dealing with diversity in the classroom.

The study concludes that despite progress in research on inclusion, significant gaps remain, particularly regarding the practical implementation of inclusive strategies in science teaching. The article suggests that future studies should further explore teacher training and evaluation of inclusive practices in the classroom.

## METHOD

The study was carried out through systematic review methods [17], drawing on a range of sources including journal articles and conference proceedings.

This study employs the Systematic Literature Review (SLR) approach as its research method. SLR is a structured technique for gathering and analyzing secondary data to address predefined research questions. The review process in this research follows the PRISMA (Preferred Reporting Items for Systematic Reviews and Meta-Analyses) guidelines, a widely recognized framework that outlines essential components for conducting and reporting systematic reviews. PRISMA consists of four key phases—identification, screening, eligibility, and inclusion—which collectively ensure a transparent and rigorous review process [17].

*Framing questions:* What are the latest advancements and best practices in teaching physics to blind and visually impaired students, and how do these approaches enhance their understanding of physical concepts?

*Identifying relevant work:* During the identification phase, references were gathered through searches across major databases, namely ERIC [18], Google Scholar, and other sources such as the *American Journal of Physics*, *European Journal of Physics*, *The Physics Teacher*, and *Physics Education*. The total number of articles from Google Scholar and Eric databases was collected using the keyword "physics teaching" and "blind" selected from anytime, resulting in the number of articles (n = 1,770) and (n = 3), respectively. The total number of articles from those databases was collected using the keyword "visually impaired physics" selected at any time, resulting in the number of articles (n = 17,800) and (n = 11), respectively. Due to the existence of previous review papers (see previous section) and trying to fill possible gaps, one limited the time selection to papers from 2014.

*Assessing the quality of studies:* During the identification stage, a total of 19,582 articles were screened based on inclusion and exclusion criteria, as outlined in Figure 3.1, to obtain research data relevant to the learning objectives. It was found that three articles did not align with the research objectives in terms of variable suitability, reducing the dataset to 300. These remaining articles were then evaluated for eligibility to ensure they provided appropriate supporting data. Of these, 272 were deemed ineligible as their results did not include the necessary variables for this study. The outcome yielded (n = 28) papers analyzed in this review. Figure 3.1 clearly illustrates the process of data collection leading to the final dataset. Table 3.1 presents the inclusion/exclusion criteria used during this review.

*Summarizing the evidence*: Data synthesis consisted of tabulation of study characteristics, quality, and effects, as shown in Table 3.2.

*Coding Scheme Employed:* The categories used in this review were the subjects: mechanics, oscillations and waves, optics, modern physics, and electromagnetism. There are eight categories that were used to understand each article in the review process, specifically on content analysis: mechanics, electricity, optics, sound, waves, thermal physics, astronomy, and general topics (methods, strategies, materials…).

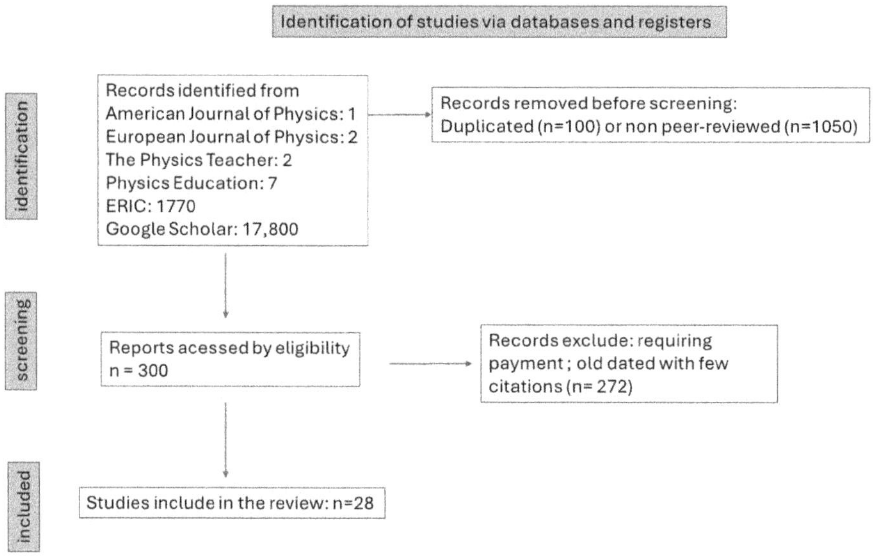


FIGURE 3.1 PRISMA steps.

TABLE 3.1 Inclusion and exclusion criteria for the reviewed papers

| | |
|---|---|
| **Inclusion criteria** | Literature review in the form of journal articles<br>Previous literature reviews<br>Articles relevant to the research topic,<br>Open access (free)<br>Articles published from 2014 to 2024<br>Articles published in English, Spanish or Portuguese<br>Peer review<br>Papers on new materials<br>Papers on adaptations of existing materials<br>Gaps in previous reviews |
| **Exclusion criteria** | Journal articles unrelated to the topic<br>Not open access or requiring payment<br>Articles not published in English, Spanish or Portuguese<br>Non peer review<br>Papers before 2014 |

TABLE 3.2 Information on scientific productions found in the survey

| Topics | References | Segment |
|---|---|---|
| Mechanics | • Bulbul et al. [20]<br>• Lisboa and Peña [19] | College |
| | • Gonçalves et al. [21] | High School |
| Electricity | • Velloso et al. [7]<br>• Velloso et al. [8] | College |
| | • Okcu and Sozbilir [23]<br>• França and Siqueira [14]<br>• Gabriel et al. [24]<br>• Stein et al. [25] | High-School |
| Optics | • de Azevedo [27]<br>• Azevedo and Santos [28]<br>• Buzzá et al. [26] | High-School |
| Sound | • Toenders et al. [29] | High-School |
| Waves | • Lisboa et al. [30] | College |
| | • Silva and Santos [31]<br>• Silveira et al. [32] | High-School |
| Thermal Physics | • Negrete et al. [34]<br>• Cordova et al. [33] | College |
| Astronomy | • Almeida et al. [35]<br>• Azevedo et al. [36] | High-School |
| General topics | • Holt et al. [37]<br>• Toenders et al. [38]<br>• Torres and Mendes [39]<br>• Dickman et al. [40]<br>• Bülbül et al. [41]<br>• Supalo et al. [42]<br>• Supalo et al. [43]<br>• Maguvhe [44] | High-School |

## RESULTS

Table 3.2 presents the reviewed papers and their categorization according to the content: mechanics, electricity, optics, sound, waves, thermal physics, astronomy, and general topics.

## MECHANICS

Lisboa and Peña [19] detail the development and implementation of a laboratory activity designed for blind students to measure the inertia of simple objects, specifically a disc around its axis of symmetry. The adaptation involved modifying the data collection process by utilizing an open-source electronic platform to translate visual signals into acoustic signals. This innovation enabled a blind student at their university to fully engage in the mechanics laboratory session of a standard engineering course alongside their peers.

Lisboa and Peña [19] developed, manufactured, and calibrated a physics laboratory experiment tailored for blind students, enabling them to measure the moment of inertia of an acrylic disc. The primary goal of this inclusive design was to allow the student to independently handle equipment and instruments, control variables, and collect experimental data. This activity replicates a traditional physics lab experiment commonly used in science and engineering programs, where measurements are typically obtained visually.

The authors designed a device that measures angular accelerations and provides auditory feedback for the results. The setup included a slotted disk mounted near the perimeter of a bronze cylinder that rotated with minimal friction around its axis. The acrylic disc, whose moment of inertia was to be measured, was positioned concentrically on the cylinder. A photogate sensor detected the slots on the disk, sending signals to an interface that calculated the angular acceleration. The measured value was then vocalized through a small speaker at the press of a button. An additional sensor emitted a "beep-beep" alarm to indicate when the system was in motion.

According to the authors [19], the blind student successfully completed the laboratory experiment independently. They positioned the disc correctly on the bronze cylinder, carefully wound the string to avoid collisions with the pulley, released the disc for free rotation, and pressed the buttons to take measurements at the appropriate moments. After auditing and recording the data, the student stopped the disc, added extra mass to the hanging weight, and repeated the process.

Bulbul et al. [20] present a study investigating whether blind students conceptualize force and motion differently from their sighted peers. The research specifically examines whether visual experiences influence students' understanding or misconceptions of these concepts. Conducted as a case study, the research gathered data from six blind high school students through conceptual physics problems related to force and motion. The analysis indicated that while the blind students' understanding was not drastically different from that of sighted students, certain conceptual difficulties appeared to be unique to them due to their lack of visual experiences. Overall, the findings suggest that visual experience does not play a significant role in shaping conceptualizations of force and motion.

The authors' findings indicate that despite their inability to make visual observations, visually impaired students perceive the concept of force similarly to their sighted peers. While both groups rely on different senses to observe the world, they provide comparable explanations when responding to questions about force. One limitation of the study is the small sample size of six participants, as larger studies could yield more reliable results. Additionally, since all participants were female, including both male and female students in future research could help determine if gender differences exist. However, the findings suggest that both blind and sighted students share similar misconceptions, despite acquiring knowledge through different sources.

This study, along with previous research on sighted students' misconceptions based on Force Concept Inventory (FCI) results, indicates that blind and sighted students reach the same conceptualization level without formal instruction. Therefore, structured instruction plays a crucial role in helping blind students achieve the same level of understanding as their sighted peers.

While sighted students may rely on visual experiences, blind students develop their conceptual understanding through social interactions, using their peers' explanations rather than personal visual observations. This indirect form of experience does not hinder their conceptual development. The authors conclude that to ensure student success, educators should first assess their pre-instruction knowledge and then design appropriate instructional materials. Additionally, peer support and collaborative learning should be integrated into all lessons to enhance their learning experience.

In response to the challenge of teaching physics to visually impaired students, Gonçalves et al. [21] present a method for demonstrating the relationship between distance and time in a pendulum experiment for blind learners. Using an Arduino and an ultrasonic sensor, the periodic oscillation of the pendulum is converted into a corresponding variation in frequency emitted by a speaker. The key advantage of this approach is that it allows blind students to perceive the motion without needing to physically touch the pendulum. The position range was converted into a corresponding frequency range, played through a computer speaker, enabling visually impaired students to perceive the pendulum's periodic motion.

## ELECTRICITY

Velloso et al. [8, 22] proposed an approach to adapting an Introductory Electricity and Magnetism Laboratory course for a blind Engineering student during the COVID-19 pandemics. They developed a pilot project aimed at bridging these conceptual gaps and enabling the student to engage with the course material effectively.

Due to the social isolation requirements of the COVID-19 pandemic, all classes were conducted remotely, and all instructional materials had to be designed and delivered to the student's home with these constraints in mind. The activities, teaching methods, and materials developed specifically for that project proved to be highly effective. They enabled the visually impaired student to participate in key experimental procedures in a manner closely resembling that of his peers. This included assembling circuits, using multimeters, marking data points for graph construction, and calculating electrical resistance from the slope of the plotted curve.

Okcu and Sozbilir [23] aimed to effectively teach eighth-grade middle school students with visual impairments about the transformation of electrical energy into light energy. Before designing the instructional materials, the specific needs of these students were identified, considering both their general and individual requirements. A case study approach was employed as research design, with seven visually impaired students participating in the study.

Castro et al. [24] describe a simple and easy-to-perform experiment designed to teach the concept of Ohm's first law to visually impaired people, using low-cost and easily accessible materials. Through the device, it is possible to understand that the intensity of the electric current, inversely proportional to the value of the circuit's resistance, is determined by the sound emitted by a buzzer. The results show three sound intensities according to the value of the resistance used in the circuit.

The authors claim that to have a satisfactory inclusion of students with visual impairment in physics classes, it is necessary to offer activities that stimulate observation, investigation, and experimentation for the development of tactile and auditory perceptions, effectively contributing to the process of abstraction and generalization of knowledge. Their proposal aims to include the student with visual impairment as an active subject, by providing an interpretation of the stimulus. The student observes, manipulates, and analyzes the experiment with the mediation of a physics teacher who will provide all the theoretical basis on the subject. The activity can also be worked on in regular classes, which favors the understanding of both visually impaired and sighted students.

Stein et al. [25] discuss that the immense technological advances experienced in recent decades come essentially from the development of devices that employ electronic circuits in their design. Students have their first contact with this topic in the physics curriculum component in high school. Even though these devices are used on a large scale, the approach to this topic is not trivial. However, when it comes to teaching the association of capacitors and resistors to people with visual impairments, the situation becomes extremely challenging, as there are still no didactic tools to support the teaching-learning process for this audience. Based on this panorama, the authors present the development of two didactic three-dimensional circuit puzzles built by 3D printer, which allow for playful learning in the study of series, parallel and mixed associations of capacitors and resistors. In addition, the prototypes were designed to meet the needs of students with visual impairment.

The tactile-visual technology presented by the authors consists of two three-dimensional educational puzzles that can be used in teaching capacitors and resistor circuits to illustrate series, parallel, and mixed associations. These prototypes can be used in the teaching-learning process for sighted students and those with visual impairments, since the pieces contain information in Braille to meet the needs of this audience. The strategy of introducing puzzle assembly in circuit teaching seeks to provide meaningful and contextualized learning, considering that the student will assemble and handle in practice the various diagrams that are presented in the books. The practical and conceptual exposition presented in their proposal provides a playful tool for teaching physics and a resource for including people with visual impairments in regular education.

## OPTICS

Buzzá et al. [26] discuss that despite being provided for by law, access to quality education for people with disabilities still faces major obstacles, especially in physics education, where the expository nature of classes and strong dependence on graphic elements predominate. The objective of this work was to develop alternative approaches that facilitate the learning of physics concepts for people with visual impairments, as well as for sighted people. Twenty-seven panels were constructed, which constituted the exhibition "Light at your fingertips," with tactile, visual, and auditory resources to explain optical concepts and phenomena, in addition to demonstrating to teachers the possibility of developing complete and inclusive material for teaching physics. Addressing topics ranging from geometric optics to electromagnetic waves and wave-particle duality, each of the panels was implemented with explanatory texts in graphic form, in braille and in the form of audio description, adapted to contain the location of the tactile elements on the panels.

Azevedo et al. [27] outline a teaching method for explaining light propagation, reflection, and refraction to visually impaired students using laser light. The approach is based on the idea that teaching physical phenomena should be supported by the creation of conceptual models. Since the theoretical framework of modeling involves procedural knowledge, it is most effectively learned through specific modeling activities, where the theory is introduced progressively to guide and monitor these activities. Therefore, hands-on tasks aimed at constructing models of physical reality should be effective in helping students develop a stronger physical intuition.

Azevedo and Santos [28] examine the challenges faced by visually impaired students when working with graphics and diagrams in their study of geometrical optics. The case study proposes practices that utilize low-cost, easily accessible, and manageable materials, offering tactile experiences for visually impaired students. The activities make use of light, portable magnetic boards and various types of magnets, primarily rubber magnet strips. A student-centered learning approach, specifically Karplus's learning circles, is implemented in this method.

## SOUND

Toenders et al. [29] present a component of a broader project aimed at developing and accessing a method for teaching an entire physics topic to visually impaired students, incorporating the student, educators, and the publisher of educational materials. In this article, we outline the creation and use of an adapted braille textbook, tactile drawings, and 3D-printed models to teach the physics of sound to a visually impaired student. This approach was recommended by the physics teacher participating in the research, who identified the numerous graphs in the study of sound as a challenge when teaching blind students. The study demonstrated that redesigning and supplementing braille study materials enhanced the learning experience and made physics more accessible for visually impaired students in mainstream high schools. By removing irrelevant content, modifying illustrations, and incorporating 3D models, greater insight was gained into how physics can be taught to visually impaired students.

Challenges in teaching physics to visually impaired students were found not only in the materials themselves, such as inaccessible drawings or references to images and graphs, but also in the teacher's time for lesson preparation and instruction, as well as the social context in which the student learns. Adapting the braille textbook led to improvements in learning physics, though some limitations were also identified. The materials cannot be overly modified, as this might prevent the visually impaired student from staying in sync with their peers.

## WAVES

Lisboa et al. [30] developed a Physics Teaching Lab experiment designed for blind students to measure the wavelength of standing waves on a string. The adaptation involved replacing the traditional visual inspection of nodes and antinodes with an alternative method based on sound volume. By using a guitar pickup to detect variations in sound at different points along the string, blind students can determine the wavelength of the standing wave. This modification enables a blind student at our university to actively participate alongside their classmates in the laboratory session for the wave unit of a standard engineering course. The goal was to enable a blind student to fully engage in the measurement process, beyond their usual participation in data analysis and report writing. Instead of relying on visual inspection of nodes and antinodes, our approach used variations in sound volume detected by a guitar pickup at different points along the string. To determine the antinode positions—where the most significant sound differences occur, the authors integrated a distance measurement system using a position sensor connected to an Arduino platform. The combination of these tools allowed the student to accurately measure half-wavelengths, achieving results within the expected error range with ease.

Silva and Santos [31] presented the development and testing of an educational tool titled *High-Relief Sheets for Teaching Waves and Wave Phenomena to Visually Impaired Students.* A total of 47 high-relief sheets were created using plastic material and commercial punches, including a 5 mm center punch for image contours, a PZ1 punch for curve representation, and a T10 punch to emphasize distinct points or structures. The content covered fundamental wave concepts typically found in high school textbooks. The product was tested in over ten classes, each lasting one hour and forty minutes, with a fully blind student. The student's prior knowledge was assessed through a pre-test featuring questions from the National (Brazilian) High School Examination. Following the lessons, conceptual understanding was evaluated using a post-test, which included five questions from the pre-test and five additional National Examination questions. Based on the students' post-test performance and their evident satisfaction, expressed through various comments, the product is considered to have strong potential for large-scale implementation.

Silveira et al. [32] discuss that the models used to teach waves, particularly at the high school level, often fail to align with the actual cognitive development of most students. Additionally, the challenge of demonstrating frequencies, whether electromagnetic or sound, beyond the range perceivable by human senses creates a significant epistemological barrier. To address this issue, the authors developed two experiments that explore the relationship between invisible and inaudible frequencies. By processing these frequencies using an Arduino platform, they are converted into audible sounds. This approach aims to make the invisible world more accessible, allowing all students, including those with blindness or deafness, to engage with the experiments in the classroom.

The purpose of the experiments developed by the authors is to provide students with the means to experience frequency bands that our senses cannot detect, both inaudible and invisible. In this way, they will be able to accept more naturally that there is a world that is undetectable to our senses, and that, for it to be perceived, it is necessary to develop physical models that relate, using analogies, physical phenomena of our daily lives, such as waves in water that are related to electromagnetic waves. Evolutionary characteristics will also be addressed, such as the fact that, in nature, some animals had their senses developed differently from those of human beings and that, therefore, they have vision and hearing limits, in some cases, beyond our perception.

## THERMAL PHYSICS

Cordova et al. [33] detail the construction of a thermometer that provides temperature readings through audio output. Originally designed to assist visually impaired students in thermometry experiments, the device could be adapted for measuring various other physical quantities. The project utilizes an Arduino Uno board along with an Arduino MP3 shield for audio playback. Temperature measurements are taken using the DS18B20 digital sensor, which operates within a range of –55 to 125°C. The entire system is simple to assemble, with readily available and affordable components.

The prototype of the audio thermometer has proven to be very efficient and robust, enabling classroom applications, specifically targeting mixed-gender classes that include visually impaired students. The authors point out that the platform can be easily adapted to other sensors. There is a wide variety of sensors for Arduino that allow us to foresee the construction of an audio-voltmeter, audio-ammeter, etc.

Negrete et al. [34] developed a teaching laboratory experiment for visually impaired students to measure the linear thermal expansion coefficient of an object. Using an open-source electronic prototyping platform, the authors created interactive electronic devices that convert visual signals into auditory cues. The goal was to allow a visually impaired student to fully engage in the measurement process, beyond their usual role in data analysis and reporting. The modification involved replacing the traditional dial-based length gauge with a segmented disk connected to a photogate sensor, instead of the standard needle on the dial's axis. The setup allowed the sensor to detect the disk's rotation angle, which corresponds to the expansion of the material—an aluminum tube in this case.

The photogate sensor data was processed using an Arduino platform, programmed to count the number of slits passing through the sensor and convert this information into audible signals via a speaker. The authors tested the device in a real lab session with a blind student. The student was able to perform the measurements independently, achieving results comparable to those of other student groups who completed the experiment successfully. Additionally, the student expressed great satisfaction in being actively involved in the measurement process, beyond the typical data analysis and reporting tasks.

## ASTRONOMY

Almeida et al. [35] focus on the development of a didactic kit utilizing touch and thermal sensors to teach Astronomy to visually impaired students. The idea emerged from the challenges of finding effective experiments that clearly convey key astronomical concepts to individuals with visual impairments. The kit features a model with hemispheres equipped with Peltier pellets to simulate the temperature scale of planets in the Solar System. The research prioritized the use of accessible and low-cost materials. This paper details the materials, equipment assembly, and the entire design process leading to the final application for students.

Azevedo et al. [36] defend that with the process of including visually impaired students in regular school systems, new alternatives must emerge so that these students not only have the right to enroll, but also to actively participate in the teaching-learning process. Physics teaching must lose its purely visual aspect to include all students in the classroom. Given this context, the authors present an inclusive pedagogical method to be used in physics classes, addressing concepts related to geocentrism and heliocentrism, the content of the first two months of classes in the first year of high school in the state of Rio de Janeiro. The material developed consists of two tactile-visual models built with low-cost material. The use of this material presented a satisfactory result, as it benefited the teaching-learning process of students with and without visual impairment, making the content covered more concrete and closer to reality.

## GENERAL TOPICS

Holt et al. [37] address the responsibilities of instructors in ensuring accessibility, including making class sessions, curricular materials, tutorials, and more inclusive demonstrations. It distinguishes these efforts from administrative accommodation, such as extended test time. The authors present resources and practical guidance for implementing these strategies, along with a recommended reading guide.

According to the authors, a blind student attending mainstream class sessions should have access to one or more individuals who serve as both an in-class assistant and an out-of-class tutor. The in-class assistant ensures that all course materials are accessible in real time while also fulfilling the typical responsibilities of a tutor when needed.

The authors suggest that classroom selection should take several factors into account. Poor acoustics or background noise from hallways or ventilation systems can make it challenging for students who primarily rely on hearing to concentrate. Seating arrangements should allow the student and their assistant to communicate quietly without disrupting the class. To accommodate assistive technologies, the student should have access to a table-and-chair setup rather than a tablet armchair. Additionally, since the student needs time to plan their route to class, any changes in location should be communicated as early as possible.

In addition, the authors suggest that all visually based course materials must be converted into an accessible format, either auditory or tactile, depending on what best suits the student and the material. Various printers and devices can produce tactile graphics, which use raised dots and other textures to represent images, figures, and diagrams from textbooks and other instructional resources. Additionally, several programs and tools can scan and convert printed text into synthetic speech or tactile output.

Holt et al. [37] also identified accessible classroom demonstrations and laboratory exercises. For instance, tactile electronic circuits can be constructed using Snap Circuits Jr. Access Kits, which include components labeled in Braille. Some experiments in oscillations and waves are inherently tactile and/or audible. For example, students can locate the resonance of vibrating strings by listening to and feeling the vibrations while varying the driving frequency. Indeed, many physical phenomena cannot be directly sensed visually; all students must use detectors to measure the inverse-square law, absorption of radiation by materials, atomic emission spectra, and similar concepts.

Toenders et al. [38] assessed the physics learning environment of a blind high school student in a mainstream Dutch school. They observed that visually impaired students to understand physics concepts, additional time and supportive materials appear to be crucial. However, teachers often have limited time to develop effective teaching strategies for these students. The authors provide recommendations for improving both the learning environment and the materials used.

The interview with the physics teacher revealed that the school does not allocate additional time for teachers when accommodating a visually impaired student in regular classes. However, teaching such a student requires adapting materials to suit both the individual and the subject. Additionally, significant one-on-one time is needed during lessons to ensure the student fully understands the concepts. It is essential for school management to inform teachers about the available resources for supporting visually impaired students. Providing teachers with sufficient time to develop and acquire appropriate materials could greatly enhance physics learning.

The modified textbook contained a substantial amount of irrelevant content for the visually impaired student, including inaccessible descriptions, references to figures, formulas, tables, and impractical assignments and experiments. A revised version, designed by a ***braille-literate physics teacher, is necessary to streamline the content and make it suitable for visually impaired learners, particularly when the textbook is intended to be self-explanatory.

Visual elements such as pictures and graphs can be made accessible by converting them into relief drawings or 3D models that effectively communicate the intended concepts. If these supplementary materials are to be self-explanatory, an experienced physics teacher should be involved in their design to ensure clarity. Additionally, the study's teacher discovered—almost too late—that a relief drawing board could be a valuable teaching aid.

Considering all these aspects, a complete redesign of the physics learning environment would be necessary to effectively teach a visually impaired student. However, since few visually impaired students pursue physics at this school, the cost of such an adaptation may not be justified for school management. If more visually impaired students were interested in studying physics, an adapted learning environment would be a logical starting point.

Over the course of the school year, the students in this study developed a more negative attitude toward physics and mathematics, finding them difficult and unengaging. Her physics grades were significantly lower compared to other subjects. During a discussion at the Blind Science Conference, another physics teacher shared experiences teaching a gifted visually impaired student in a regular high school, highlighting different challenges than those encountered in this study.

While adapting study materials may improve a visually impaired student's understanding of physics, it has been suggested that these students still require additional time to process new information. This is because touch and sound are processed sequentially in the human brain, unlike sight and sound, which are processed simultaneously. Another crucial factor influencing physics learning is the age at which the student became visually impaired, as well as the degree of vision loss.

Torres and Mendes [39] developed a didactic kit called KitFis. It emerged from qualitative research aimed at developing and accessing educational materials capable of tactilely reproducing the illustrations used in physics instruction. The kit is constructed from durable materials, ensuring its longevity even with frequent use by students and teachers. Its ability to reproduce multiple illustrations makes KitFis a versatile teaching tool. Additionally, the illustrations can be easily deconstructed and reassembled as often as needed. This feature is particularly beneficial for teachers, as it reduces the time required for planning and preparing supplementary materials in advance.

Dickman et al. [40] designed and evaluated a set of tactile symbols to represent commonly used elements in mechanics diagrams, such as vectors, ropes, pulleys, blocks, and surfaces. These symbols enable the adaptation of physics illustrations in textbooks for blind students. Additionally, we explored how figures are described to blind students during classroom activities and exams by conducting interviews with three blind students using the oral history method. The symbols were tested at a specialized school for the blind, and our findings suggest that, with training, blind students can recognize and interpret the symbols in a problem without requiring a verbal explanation. This educational tool promotes greater autonomy for blind students, allowing them to study physics under conditions comparable to their sighted peers.

During the test, participants were able to identify the elements in a physics diagram and correctly answer the given question by manipulating the tactile symbols and listening to the oral description of the scenario. The results suggest that, with adequate training, blind students could independently recognize these symbols in a problem without requiring a verbal explanation.

Bülbül et al. [41] suggested adapting smart phone applications about physics education to blind students. According to the authors, their study serves as a preliminary survey for instructional design, examining free smartphone applications related to physics using English keywords in the Google Play Store. The apps were analyzed based on their star ratings and number of downloads. We then evaluated their accessibility for blind users and explored ways to create scenarios that would facilitate their use for both sighted and visually impaired individuals.

Supalo et al. [42, 43] analyzed data from a multi-year field study to assess whether a set of audible adaptive technologies could enhance hands-on participation for high school students with blindness or low vision in chemistry and physics courses. The study involved students in mainstream classrooms across the United States, working alongside sighted laboratory partners. Four data categories were examined to evaluate hands-on participation: quantitative coding of video-recorded lab lessons, qualitative assessments of the same videos, and interviews with both students and teachers. The findings provided strong evidence that these technologies effectively increased students' hands-on engagement in laboratory activities. However, certain factors influenced quantitative analysis. Students with some usable vision showed similar participation levels with and without the adaptations, while those with minimal vision often required more time to complete certain lab tasks. Additionally, various elements inherent to real-world educational settings were found to significantly impact student outcomes.

The authors found that students with visual impairment often needed more time than their sighted peers to complete certain laboratory tasks. It suggests that future research should integrate quantitative measurements with an analysis of time on tasks. This approach would provide a more comprehensive metric that accounts for both the duration and the effectiveness of time use. Additionally, improving control over various influencing factors and providing more thorough training on experimental technologies before interventions would be beneficial for future studies.

As adaptive technologies and teaching methods continue to evolve, ongoing research should assess their effectiveness using both quantitative and qualitative measures. Further studies should also examine a range of educational outcomes, including learning gains and academic achievement. The goal is to foster continuous innovation in laboratory adaptations for visual impairment students, ensuring greater accessibility to science education and promoting educational equity.

The authors conclude that hands-on chemistry education for visually impaired students is still in its early stages, and this study represents just one example of what is possible. Further contributions are needed to address challenges in general chemistry and even more so in advanced courses. The technologies explored in this study were designed to empower visually impaired students, demonstrating that they can actively contribute to laboratory work. This research also aims to inspire chemistry educators to raise their expectations and create more inclusive science learning environments. Much work remains to ensure that chemistry and other scientific disciplines are as accessible and inclusive as possible.

Maguvhe [44] examined the factors that hinder the participation of blind and partially sighted learners in mathematics and science education. Given that teachers remain one of the most significant influences in any education system, the researcher considered it important to explore the teacher's role from the perspective of a blind technician in encouraging the involvement of blind and partially sighted learners in these subjects, which many of these learners do not pursue beyond primary school. A case study was conducted with a blind technician, who identifies as an unqualified scientist, to understand how various school-related factors might encourage blind and partially sighted students to engage in mathematics and science education, as well as help retain them in these fields. The participant shared insights from his personal experiences within the school environment and broader social institutions. A semi-structured interview format was used, and the responses were recorded with mutual consent. Analysis was based on the questions posed during the interview. The study found that there is a lack of teacher motivation and mentorship in mathematics and science methods, as well as insufficient use of tools for empowering students. It also highlighted that teachers lack the necessary skills in special education to fully support learners' potential in these subjects. These issues point to the need for government intervention in teacher training and development.

Maguvhe [44] attempted to determine the factors affecting the participation of blind and partially sighted learners in science and mathematics in schools for the blind. From the findings of the study, it can be concluded that successful mathematics and science education for blind and partially sighted learners requires teacher empowerment through rigorous training and development on both approaches to subject matter and general awareness to the unique learning needs of learners with various degrees of visual loss. It must be noted that, since this was a small case study, the chance of many teachers and former learners expressing divergent views exists, although the findings corroborate the results of similar studies discussed in the literature review. Although case studies are based on such small samples that their results are less generalizable, experiences of people in special populations might have some commonalities. This study could inspire the quest to conduct quantitative studies in several African countries to understand the perceptions of former learners as products of the teaching/learning process to determine the needs of current regional education systems.

## CONCLUSION

This chapter emphasizes the importance of further exploring how the various elements of different disciplines can be made accessible to all students by considering their sensory abilities. It also serves as a foundation for future research aimed at advancing understanding of the cognitive processes through which students with visual functional diversity develop scientific knowledge in educational settings. The next chapter presents a detailed discussion of how to build adapted material and teaching strategies to support the teaching of the visually impaired student.

It has been suggested that students with visual impairments can effectively learn science when provided with appropriate supportive tools. These tools may include resources for orientation and mobility, tactile and kinesthetic learning aids, auditory learning accommodation, and assistive technologies tailored to the student's level of vision. With the right support, students with visual impairments have the potential to excel in science education.

Students with visual impairments are capable of excelling in physics. This book highlights the importance of conducting research on science learning approaches for students with visual impairments to evaluate their impact on student achievement. A student-centered learning model incorporating assistive technologies is recommended, along with classroom settings that are specifically adapted to meet the needs of students with visual impairments.

Individuals with visual impairments have been marginalized for a long period of time, often viewed as a burden. It is now time for physics educators to prioritize the knowledge, skills, and experiences of disabled individuals and to advocate for systemic changes that enable their full participation in physics classrooms.

These works underscore the effectiveness of adapted resources in fostering inclusion and enhancing understanding for students with disabilities. They also demonstrate the diverse approaches—ranging from sound-based methodologies to tactile models—that make physics education accessible. Authors repeatedly emphasized the importance of these innovations in engaging students, building autonomy, and showing that physics education need not rely solely on visual representation.

Supalo et al. [43] explored the collaboration between science faculty, teachers of visually impaired students, the students themselves, and their parents or legal guardians. By working together, these key stakeholders can leverage their strengths and knowledge to create a more effective implementation plan, ensuring students with visual impairments have a successful hands-on science learning experience.

## BIBLIOGRAPHY

1. Castro-Castillo, D. C., & Tuay-Sigua, R. N. (2023). Revisión sobre la enseñanza de la física en estudiantes con diversidad funcional visual. *Revista científica, 46*, 1–16.
2. Camargo, E. (2008). *Ensino de física e defiencia visual: Dez anos de investigacioes no Brasil.* São Paulo: Plêiade/FAPESP.
3. Camargo, E., & Nardi, R. (2013). Contextos comunicacionales adecuados e inadecuados para la inclusión de alumnos con discapacidad visual en clases de física moderna. *Enseñanza de las ciencias, 31*(3), 155–175.
4. Camargo, E., Nardi, R., & Peralta, D. (2010). Contextos comunicacionais adequados e inadequados à inclusão de alunos com defi ciência visual em aulas de termologia. *Revista de Enseñanza de la Física, 23*(1–2), 21–40.
5. Camargo, E., Nardi, R., & Rodrigues, E. (2012). Análisis del proceso inclusivo del alumno ciego en clase de física moderna. *Góndola, 7*(1), 6–31.
6. Camargo, E., Scalvi, L., & Braga, T. (2007). Concepciones alternativas sobre reposo y movimiento, modelos históricos y deficiencia visual. *Enseñanza de las ciencias, 25*(2), 171–182.
7. Velloso, M., Arana, M., Acioly, V., & Santos, A. (2021). Hands-on electricity remote teaching to a blind student during pandemic of 2020. *Physics Education, 56*(5), e10.
8. Velloso, M., Arana, M., Acioly, V., & Santos, A. C. F. (2021). Abordagem da Lei de Ohm para Estudantes com Deficiência Visual no Ensino Superior. *Latin-American Journal of Physics Education, 15*(1), 10.

9. Lisboa, A., Peña, F., Negrete, O., & Dib, C. (2021). Teaching labs for blind students: equipment to measure standing waves on a string. *European Journal of Physics, 42*(6), e18.
10. Ediyanto, & Kawai, N. (2019, June). Science learning for students with visually impaired: a literature review. *Journal of Physics: Conference Series, 1227*(1), 012035. IOP Publishing.
11. Chini, J. J., & Scanlon, E. M. (2023). Teaching physics with disabled learners: a review of the literature. *The International Handbook of Physics Education Research: Special Topics*, 1-1. Melville, NY: AIP Publishing.
12. Pacala, F. A. (2024). The use of technology in teaching physics to students with disabilities: a systematic review. *Information Technologies and Learning Tools, 104*(6), 53.
13. da Silveira, M. V., & dos Santos, A. C. F. (2022). A percepção dos professores de Física sobre a inclusão de estudantes com deficiência visual: uma pesquisa quantitativa. *Revista Educação Especial, 35*, 1–31.
14. França, S., & Siqueira, M. (2019). Propostas didáticas no ensino de física para deficientes visuais: análise de trabalhos em periódicos e eventos nacionais (2000–2018). *Latin-American Journal of Physics Education, 13*(4), 3.
15. de Oliveira, M. H. B., da Silva Leite, M. L., & Rafael, F. J. (2022). Levantamento bibliográfico sobre o ensino de Física inclusivo para alunos com deficiência. *Ensino em Perspectivas, 3*(1), 1–12.
16. de Souza Mól, G., de Morais, A. V., da Silva, W. P., & de Camargo, E. P. (2020). Panorama da Inclusão no Ensino de Ciências de acordo com publicações mais relevantes da Área. *Revista da Sociedade Brasileira de Ensino de Química, 1*(1), e012004–e012004.
17. Khan, K. S., Kunz, R., Kleijnen, J., & Antes, G. (2003). Five steps to conduct a systematic review. *Journal of the Royal Society of Medicine, 96*(3), 118–121.
18. ERIC database, Retrieved from https://eric.ed.gov/
19. Lisboa, A., & Peña, F. J. (2024). Teaching labs for blind students: equipment to measure the inertia of simple objects. *American Journal of Physics, 92*(11), 889–891.
20. Bulbul, M. S., Garip, B., & Özdemir, Ö. F. (2015). Using a force concept inventory test with visually impaired and blind students. *European Journal of Physics Education, 6*(3), 20–31.
21. Gonçalves, A. M. B., Cena, C. R., Alves, D. C. B., Errobidart, N. C. G., Jardim, M. I. A., & Queiros, W. P. (2017). Simple pendulum for blind students. *Physics Education, 52*(5), 053002.
22. Velloso, M., Arana, M., Acioly, V., & Santos, A. C. F. (2021). Hands-on electricity remote teaching to a blind student during pandemic of 2020. *Physics Education, 56*(5), 055027.
23. Okcu, B., & Sozbilir, M. (2019). Designing a bulb to teach electric circuits to visually impaired students. *The Physics Teacher, 57*(2), 99–101.
24. Castro, Gabriel R., Leite, Gabriela S., & Martins, Jefferson S. (2021). Construção de um experimento para o ensino-aprendizagem da primeira lei de Ohm para estudantes com deficiência visual. *Física na Escola, 19*(1), 28.

25. Stein, Cléver Reis, Santo, Artur Vitório Andrade, & Junior, Moacy José Stoffes (2021). Tecnologia tátil-visual para o ensino de associação de capacitores e resistores. *A Física na Escola, 19*(2), 123.
26. Buzzá, H. H., Campos, C. P., Requena, M. B., Andrade, C. T., Leite, I. S., Fortunato, T. C., ... & Kurachi, C. (2018). Preparação de material tátil-visual torna o ensino dos conceitos de óptica acessível para pessoas com deficiência visual-Exposição "Luz ao Alcance das Mãos". *A Física na Escola, 16*(1), 36–42.
27. de Azevedo, A. C., Vieira, L. P., Aguiar, C. E., & Santos, A. C. F. (2014). Teaching light reflection and refraction to the blind. *Physics Education, 50*(1), 15.
28. Azevedo, A. D., & Santos, A. C. F. (2014). Teaching optics to blind pupils. *Physics Education, 49*(4), 383.
29. Toenders, F. G., de Putter-Smits, L. G., Sanders, W. T., & den Brok, P. (2017). Improving physics teaching materials on sound for visually impaired students in high school. *Physics Education, 52*(5), 055006.
30. Lisboa, A., Peña, F. J., Negrete, O., & Dib, C. O. (2021). Teaching labs for blind students: equipment to measure standing waves on a string. *European Journal of Physics, 42*(6), 065701.
31. Silva, A. C. D., & Santos, C. A. D. (2018). Lâminas em alto-relevo para ensinar fenômenos ondulatórios a deficientes visuais. *Revista Brasileira de Ensino de Física, 40*(4), e5406.
32. Silveira, M. V. D., Barthem, R. B., & Santos, A. C. D. (2018). Proposta didático experimental para o ensino inclusivo de ondas no ensino médio. *Revista Brasileira de Ensino de Física, 41*, e20180084.
33. Cordova, H. P., Aguiar, C. E., Amorim, H. S. D., Sathler, K. S. O., & Santos, A. C. F. D. (2018). Audiotermômetro: um termômetro para a inclusão de estudantes com deficiência visual. *Revista Brasileira de Ensino de Física, 40*(2), e2505.
34. Negrete, O., Lisboa, A., Peña, F. J., Dib, C. O., & Vargas, P. (2020). Teaching labs for blind students: equipment to measure the thermal expansion coefficient of a metal. *European Journal of Physics, 41*(3), 035704.
35. Almeida, M. S., Castro, J. N.M., Cruz, W. T., Almeida, R. Q. (2020). Construção de uma Maquete do Sistema Solar com Controle de Temperatura para Alunos com Deficiência Visual. *Revista Brasileira de Ensino de Física, 42*, e2019009.
36. Azevedo, S. D. S. M., Schramm, D. U. D. S., & Souza, M. D. O. (2018). Material pedagógico inclusivo: Trabalhando com maquetes tátil-visuais do modelo geocêntrico e do heliocêntrico. *Revista Física na Escola, 16*(1), 30–34.
37. Holt, M., Gillen, D., Nandlall, S. D., Setter, K., Thorman, P., Kane, S. A., ... & Supalo, C. (2019). Making physics courses accessible for blind students: strategies for course administration, class meetings, and course materials. *The Physics Teacher, 57*(2), 94–98.

38. Toenders, F. G., de Putter-Smits, L. G., Sanders, W. T., & den Brok, P. (2017). Analysing the physics learning environment of visually impaired students in high schools. *Physics Education, 52*(4), 045027.
39. Torres, J. P., & Mendes, E. G. (2017). Making visual illustrations of physics accessible to blind students. *The Physics Teacher, 55*(7), 398–400.
40. Dickman, A. G., Martins, A. O., Ferreira, A. C., & Andrade, L. M. (2014). Adapting diagrams from physics textbooks: a way to improve the autonomy of blind students. *Physics Education, 49*(5), 526.
41. Bülbül, M. Ş., Yiğit, N., & Garip, B. (2016, April). Adapting smart phone applications about physics education to blind students. *Journal of Physics: Conference Series*, 707(1), 012039. IOP Publishing.
42. Supalo, C. A., Humphrey, J. R., Mallouk, T. E., Wohlers, H. D., & Carlsen, W. S. (2016). Examining the use of adaptive technologies to increase the hands-on participation of students with blindness or low vision in secondary-school chemistry and physics. *Chemistry Education Research and Practice, 17*(4), 1174–1189.
43. Supalo, Cary A., Isaacson, Mick D., & Lombardi, Michael V. (2014). Making hands-on science learning accessible for students who are blind or have low vision. *Journal of Chemical Education, 91*(2), 195–199.
44. Maguvhe, M. (2015). Teaching science and mathematics to students with visual impairments: reflections of a visually impaired technician. *African Journal of Disability, 4*(1), 1–6.

CHAPTER 4

# Teaching mechanics

## INTRODUCTION

Teaching and learning Newtonian mechanics are challenging subjects both to people with and without visual impairment. One reason is because it demands mental representations of spatial or geometrical situations for a significant comprehension of the subject. Newtonian physics requires a shift from everyday intuition to abstract, idealized models. This abstract reasoning is especially hard for students who are used to concrete experiences. The math involved can be also a barrier, especially in calculus-based physics. Algebra and trigonometry are heavily used; vector analysis adds an extra layer of complexity.

Another barrier is that students often come with preconceptions: "Motion needs a cause," i.e., they believe constant motion requires a force; Believing heavier objects fall faster. These intuitive but incorrect ideas clash with Newtonian concepts. Students also struggle to identify and isolate all the forces acting on a body. For them, free-body diagrams can be very confusing. Distinguishing between different types of forces (normal, tension, friction, gravitational) is not obvious. The idea that forces come into *pairs* (Newton's third law) is hard to grasp, especially when the paired forces act on different objects. Newtonian mechanics assumes an inertial frame of reference, which is an abstract concept. Understanding relative motion and how it varies between observers can be confusing. Physics problems often require multi-step reasoning: interpreting the scenario, applying laws, and solving equations. Sometimes, students may know the formulas but not when or how to apply them. Finally, language and symbols:


DOI: 10.1201/9781003510079-4

A common issue in many physics classes is the failure to recognize that terms such as "force," "work," "energy," and "mass" come with rich, metaphorical associations in everyday language. When introduced in a scientific context, these words take on specialized, abstract definitions that may be completely unfamiliar and unintuitive to students. Those words have specific scientific meanings that differ from everyday use.

Blind students often engage in formal physics education without the visual experiences and representations that sighted students naturally acquire, particularly regarding physical phenomena. This lack of visual exposure can create significant barriers to understanding scientific concepts. Consequently, the gap in scientific knowledge between sighted and blind students can be substantial. To address this disparity, it is crucial to provide blind students with appropriate support. Along with this book, one demonstrates that with adequate assistance, blind students can grasp even complex without relying solely on memorization.

In the case of education for sighted students, teaching physics depends mainly on information often expressed using graphical representations, visual aids such as blackboards, slides, films, symbols, and diagrams. Additionally, two-dimensional drawings are frequently used to depict three-dimensional scenarios. Consequently, visually impaired students often find it extremely difficult to follow physics classes with medium-level content. On the other hand, literature suggests that including students with disabilities in mainstream classes (classes for sighted students) positively impacts the development of collaborative values, respect for differences, and the creation of a more inclusive society, while highlighting the diversity of human nature [1]. In addition, laws on inclusive education require schools to admit students with special needs into their regular educational programs and provide the additional support necessary to ensure their attainment. Therefore, efforts toward inclusion focus on ensuring the active participation of students with visual disabilities in activities related to teaching conceptual and procedural mechanics. This includes addressing real issues that arise in the relationships between instructors and students with visual disabilities, interactions among students with and without visual disabilities, and the engagement of students with visual impairments in mechanics knowledge, experimental activities, mathematical operations, and more [2, 3]. Thus, one must provide teaching methods and aids tailored to the perceptual abilities of the visually impaired student.

Blind students can access all necessary information through touch, given the availability of appropriate touchable objects [1–20]. Literature suggests, for instance, that visually impaired students have the same understanding of the concept of force as their sighted peers [16]. This chapter describes some examples of teaching resources specifically developed to teach physics to blind students, covering Newtonian mechanics [1–20].

## MISCONCEPTIONS

Research has shown that misconceptions in physics concepts are common among students, visually impaired or not. These misconceptions often arise from learners' attempts to link personal experiences to concepts, resulting in strongly held beliefs that differ from expert understandings. Such errors typically stem from misinterpretations of material. When physics instruction fails to address these misconceptions, students may face significant learning difficulties, leading to lower achievement levels. Uncorrected misconceptions can further solidify inaccurate beliefs and persist through higher levels of education.

The concept of force has a strong connection to everyday experiences. Bulbul et al. [21] explored how blind students understand force, by employing the Turkish adaptation Force Concept Inventory (FCI). The FCI is a widely used [21–25] assessment tool in physics education that reveals students' understanding of force. When students consistently make the same errors, these errors are identified as misconceptions, as they confidently hold incorrect beliefs.

The Bulbul et al. [21] findings reveal that, despite their inability to make visual observations, visually impaired students perceive the concept of force in much the same way as their sighted peers. The authors describe that, despite the degree of visual impairment, the students provide similar explanations to justify their responses to questions about force. The limitations of that study include its small sample size of only six participants, which restricts the generalizability of the results.

Nevertheless, Bulbul et al. [21] demonstrate that both sighted and blind students can hold similar misconceptions, even though their experiences stem from different sensory sources. The results suggest that a well-structured instructional approach enables blind students to achieve the same conceptual understanding as their sighted peers. This underscores the importance of developing inclusive curricula for physics courses.

Inclusive curricula should incorporate materials designed specifically for blind students, including tactile graphs, accessible visuals, and experimental tools. While sighted students may rely on direct visual experiences, blind students instead benefit from social interactions and explanations from peers, which effectively replace personal visual experiences. These indirect experiences do not hinder their conceptual understanding.

Bulbul et al. [21] advocate for the inclusion of blind and visually impaired students in inclusive curricula. The authors conclude that for this approach to succeed, it is essential to first assess students' pre-instruction knowledge.

## SYMBOLS AND DIAGRAMS IN HIGH RELIEF FOR THE REPRESENTATION OF DIAGRAMS IN MECHANICS

Sighted people can see multiple objects simultaneously, but the visually impaired can only explore one object at a time through touch or hearing. It is estimated that 80% of learning occurs through sight. Consequently, visually impaired students require more time to process information compared to their sighted peers.

One of the points that teachers should consider in the teaching-learning process for blind students is how to reproduce and use diagrams and symbols in high relief, constructed by dots like those in braille code, to help visually impaired students perceive, through touch, objects commonly used to illustrate problem situations in Newtonian mechanics exercises.

Science teaching relies heavily on the use of figures that illustrate the content. In the classroom, one of the main tools used by teachers is the textbook, which makes viewing the figures and texts contained therein essential for the teaching and learning process. It is known that the text is translated into Braille, but the figures that illustrate the content are left to the teacher or someone who helps the visually impaired student to study.

If images play a role in the negotiation of meanings in textbooks, it is important that alternative resources be used for those who cannot undergo this experience. Images are a widely accepted medium in scientific dialogue, with a particular potential for communicating aspects of nature and indicating the content of ideas.

Thus, if most of the images in textbooks adapted for students with visual impairments are removed, they become a tool of little use in terms of student autonomy, which is fundamental to their development.

For students with visual impairments to achieve their autonomy, it is necessary to improve the conditions of access to the content contained in books and teaching materials in the curriculum.

In this section, we present a suggestion of diagrams and symbols to represent the basic elements necessary for the construction of diagrams related to the content of mechanics.

The elements can be represented by means of symbols, formed by defined points and positioned according to a common pattern, as shown in Figure 4.1. The use of symbols simplifies the figure, thus facilitating the identification of the elements and the understanding of the problem situation by the students. In addition, it is recommended to use a few elements, however, sufficient to create a wide range of problem situations related to Newton's dynamics.

The symbols can be created using simple free software and can be used by any person or institution without the need for courses or licenses. The symbols must be well defined, facilitating standardization and assimilation by students with visual impairments. The symbol that represents each element does not necessarily have to resemble the real object, since the students are taught to read and interpret them. It is possible to use low-cost materials, making their construction feasible. Finally, the materials must be tested and approved by visually impaired students.

There are several free software programs available for creating drawings and graphs [4, 5]. For example, when representing surfaces, tables, walls, and floors, these can be represented by blocks or solid lines. The inclined plane can be represented by large solid triangles. To print diagrams, embossing printers that print in high relief can be used.

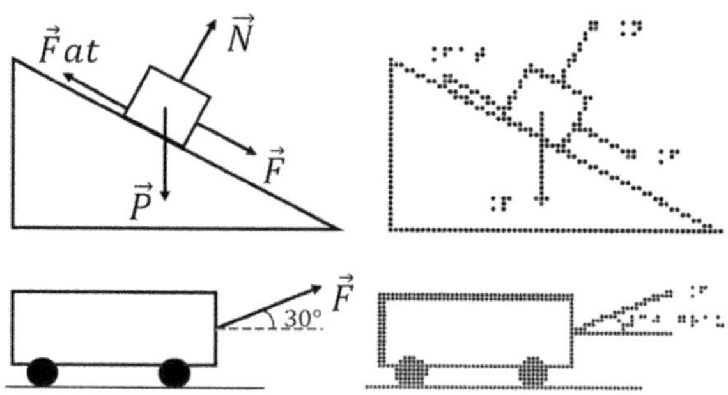


FIGURE 4.1 Top left side: Diagram of forces that act on a body on an inclined plane. Top right side: An adapted drawing using the Monet program. The forces F, N, and P are written in Braille. Bottom left side: A force that acts on a car. Bottom right side: The adapted drawing of a force acting on car. The label F and the angle with the horizontal plane are written in Braille.

Blind students typically depend more heavily on curricular materials than their sighted peers. However, textbooks and curricular materials are frequently unavailable in Braille and must be specially prepared. When materials include mathematical notation or complex diagrams, manual transcription is required. As a result, producing accessible materials in Braille and tactile graphics can take several weeks or even months.

To accommodate this, instructors should provide a comprehensive list of all material textbooks, handouts, web resources, worksheets, assignments, and lab manuals—as early as possible, ideally before the semester begins. Braille and tactile materials can be produced either through external services or in-house, depending on the scope and urgency of the task. In-house production is appropriate for brief, customized, or time-sensitive materials, such as homework or exams.

Students can access Braille and tactile graphics in various ways. Hardcopy Braille printouts provide a durable, integrated format but may be bulky and fragile. Tactile printouts, see, for instance, Figure 4.2, can be

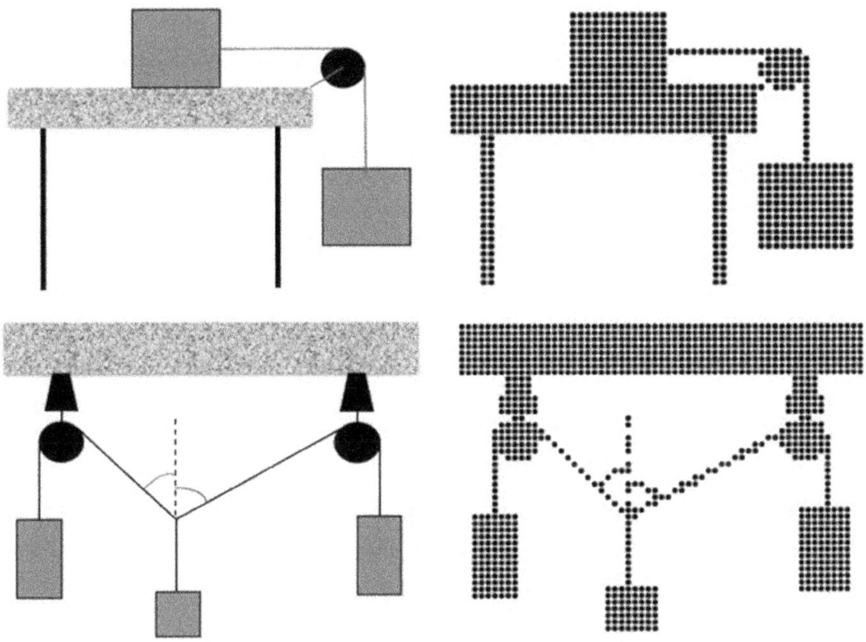

FIGURE 4.2 Examples of illustrations represented by the symbols. Left: textbook figures; right: representation.

produced using Braille embossers or thermal printers. Embossers either raise Braille dots or create tactile graphics using special heat-sensitive paper. Some models also include ink printing for sighted users.

## MATHEMATICAL WRITING FOR VISUALLY IMPAIRED STUDENTS

Understanding the physical concepts present in the most diverse phenomena around us is just as important as understanding the mathematical language that describes them. Both the concepts and the mathematics behind them are representative models of observed reality, in a complementary relationship, so that purely conceptual physics would become mathematics itself.

A major challenge when teaching physics to visually impaired students is when we introduce mathematical language. While a text is naturally represented in a linear way, through the combination of a limited number of characters, including letters, numbers, punctuation marks, and some special characters, mathematical language is represented in a two-dimensional way (superscripts, subscripts, fractions, symbols representing operations) and can contain all the elements of a text plus many special symbols, such as Greek letters, for example. This form of representation of conventional mathematical language is typical of a culture of sighted people. When solving a problem, sighted students can perceive all the variables involved with a simple scan of their eyes. For visually impaired students who use the Braille system, reading and writing is done character by character, which when simplified for sequential reading, especially fractional equations, makes it difficult to understand the whole and to recall some part of the equation [2].

With the advent of computers and the development of auditory interface tools (see, for example, [6]), it became possible for visually impaired people to access the world of computing. When we use auditory interface tools in the context of teaching physics, in terms of access to digitalized teaching materials, visually impaired students are unable to access graphic characters, especially mathematical expressions. Thus, a significant problem in teaching physics to visually impaired students lie in the complexity of conventional mathematical language.

While it is difficult to operate in Braille, on the other hand, its symbology makes it incompatible with the most varied auditory interface tools. An alternative is the LaTeX language [7]. The characters used in the LaTeX language are entirely textual, devoid of visual graphic elements, which can present barriers to screen reader software. For example, see the following equation:

$$\mathbf{F} = m\mathbf{a} \tag{4.1}$$

where the notation in bold means that it is a vector quantity, and in LaTeX, it is:

$$\backslash\text{vec}\{F\} = m * \backslash\text{vec}\{a\} \tag{4.2}$$

Something very important for a person when solving calculations or equations is the possibility of not memorizing everything while solving the problem or equation. This is exactly what a sighted person does. They read the problem or equation, write down its variables, reason, and, when necessary, review their notes with their eyes. However, Braille in its conventional form does not allow a visually impaired person to do this.

Making mathematical writing accessible to visually impaired students is essential for equity in physics. Table 4.1 shows some effective strategies and tools that support accessible math learning.

TABLE 4.1 Effective strategies and tools that support accessible math learning for visually impaired students

| Suggestion | How to implement it |
|---|---|
| Use semantic math formats | Instead of relying on visual layout (like superscripts or fractions), use formats that convey meaning through structure:<br>• **MathML**: It is a markup language for mathematics that screen readers can interpret. Mathematical Markup Language (MathML) is an XML-based language for describing mathematical notation.<br>• **LaTeX with MathJax**: Rendered math content can be made accessible using tools like MathJax with screen reader support. |
| Math Description via Text-to-Speech | Use tools that can read math aloud using standard conventions:<br>• **NVDA** or **JAWS** with **MathPlayer** (for Windows)<br>• **VoiceOver + Safari** on macOS/iOS (works with MathML)<br>• **Equatio**, which can convert written math into spoken text or Braille<br>• Encourage the use of **ClearSpeak** or **MathSpeak** styles, which describe math syntax and structure in a consistent spoken form. |

*(Continued)*

TABLE 4.1 (Continued)

| Suggestion | How to implement it |
|---|---|
| Text Descriptions and Verbal Scaffolding | Provide narrative descriptions of equations and visual math content:<br>• Explain what each symbol means and how expressions are structured.<br>• Use phrases like: "x squared" instead of $x^2$.<br>• "The square root of x" instead of $x^{1/2}$.<br>• Pair visual math with verbal walkthroughs in lectures and notes. |
| Accessible Math Input and Output Tools | Let students write and read math accessibly:<br>• Braille Math Code (Nemeth Code) for students using refreshable Braille displays.<br>• Use MathType to insert equations into Word or Google Docs—it helps generate MathML behind the scenes. |
| Accessible Document Formats | Ensure materials are in screen-reader-friendly formats:<br>• Use accessible PDFs, EPUB3, or structured Word documents with proper styles.<br>• Avoid images of equations without alt text or accompanying descriptions. |
| Involve Students in Accessibility Planning | Ask students directly:<br>• What formats work best for them?<br>• What tools or accommodations do they prefer?<br>• Do they use Braille, audio, magnification, or a combination? |
| Additional Recommendations | • Keep notation consistent.<br>• Avoid overloading symbols (e.g., don't use the same letter for multiple things).<br>• Break down complex expressions step by step.<br>• Label graphs and diagrams with textual descriptions or tactile graphics. |

Appendix A provides some more details about the writing accessible tools.

For visually impaired students who are not proficient in reading Braille, converting course materials into audio formats may offer a more effective alternative [26]. Alongside traditional options like audiobooks or having someone read materials aloud, various tools now exist to transform text into speech or Braille on demand. One such tool is JAWS (Job Access With Speech), a screen-reading software which can vocalize or convert into Braille the text output of many Windows-based applications (see Appendix A). The same company also offers MAGic, a screen magnifier designed to support individuals with low vision.

For users of Apple devices, VoiceOver, a built-in screen reader available on both iOS and macOS, functions similarly to JAWS by converting on-screen text into either speech or Braille. Some refreshable Braille displays, discussed later, also come equipped with built-in text-to-speech features, allowing users to listen to content rather than read it.

In addition, smartphone apps are being developed to offer capabilities like small-scale optical character recognition and color detection, using the device's built-in camera to make printed or visual information more accessible. When screen readers encounter images, they read aloud any available alternative text associated with the image. In HTML5 environments, such as learning management systems, this descriptive text is typically added by the content creator—often the course instructor. The same applies to documents like Microsoft Word, PowerPoint, or PDFs, where the instructor is also responsible for including accurate alt text to ensure accessibility.

## MODELS AND MULTI-SENSORY EXPERIMENTS IN MECHANICS

Graphs are commonly used in physics classes to relate two quantities, such as speed vs. time, Force vs displacement, among other relationships. How can a blind student follow the classes without being able to analyze a graph and derive from it the function that governs the phenomenon being studied? Visually impaired students have several difficulties in physics classes, graphs are one of them [8]. It has been reported that students sometimes have no idea what a graph is and are also unfamiliar with geometric figures [8]. One way to overcome those difficulties is using tactile models of geometric models [8]. For instance, a sheet of cardboard, or aluminum, or any other material, can be cut to create some geometric figures: triangles, trapezoids, rectangles, squares, and disks with different radii, which can be sensed by the students by using their hands (Figure 4.3).

Another useful tool is a geoplane, which consists of a square board with nails to build figures [8]. Figure 4.4 shows a geoplane which measures 40 cm × 40 cm, with 121 small nails arranged in 11 rows and 11 columns, 3 cm apart from each other. The first column from left to right is the vertical axis (ordinates). The first line from bottom to top is the horizontal axis (abscissa). The graphs can be drawn by rubber bands from the formation of the axes. Depending on the density of nails on the geoplane, one can make approximate graphs of curved lines, such as parabolas. These, although not perfect, can be easily understood. Figure 4.4 shows the geoplane with a simple graph: the vertical axis represents the speed of a particle, and the horizontal axis represents the time. In this exercise, the task is to determine the average speed of a particle whose speed varied with time in the way shown in the figure, that is, a rectangular trapezoid.

FIGURE 4.3 Geometric figures made of aluminum. From [8] with permission.

There are some basic principles that one should follow for preparing tactile graphics [9]. First, compose the tactile graphic as simple and clear as possible, considering the perspective of the visually impaired student. It's the teacher's responsibility to present the information in a simple, clear, and concise manner for the student. Second, the instructor should know the important information she wants to pass to the student when making the graphic. Then, the teacher should arbitrate if the shapes and textures are important or not to convey the concept. Should simple geometric shapes and braille signs be used to present the concept? Any unnecessary or irrelevant information should be omitted. It may prevent the student from gaining relevant information, therefore making the graphics useless. The graph scale should be as large and as clear as possible. One should be aware of the knowledge level, the skills, and age level of the student. The language should be appropriate to the age and knowledge level of the

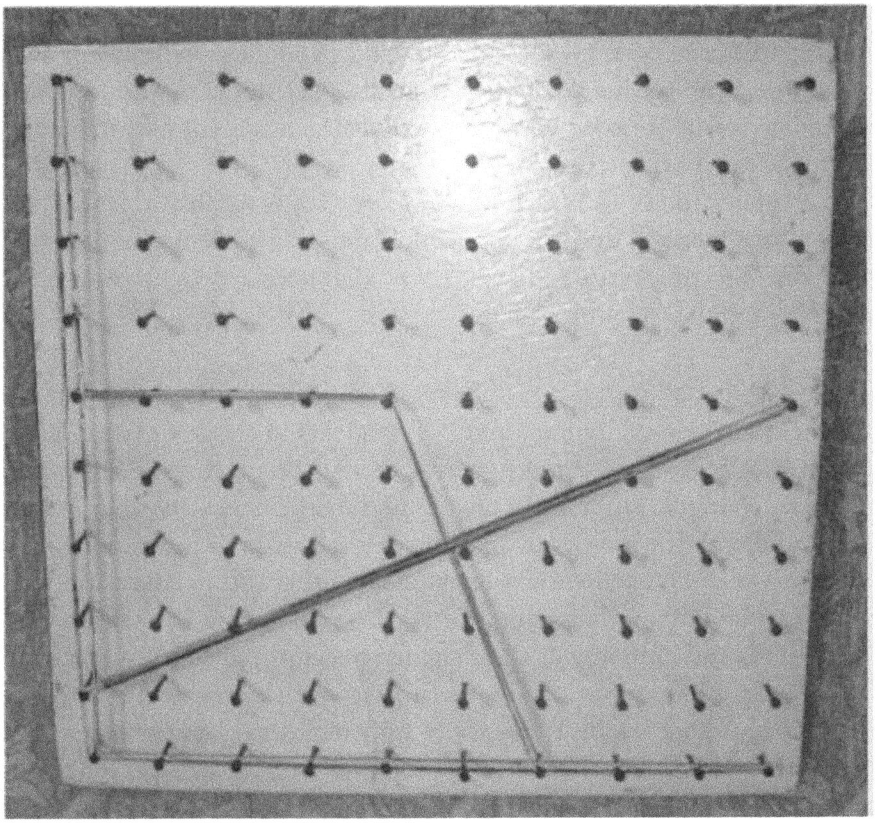

FIGURE 4.4 The geoplane. The lines are traced by drawing using rubber bands. From [8] with permission.

student. The teacher should also determine if any measurement or operation must be performed by the student. The lines and angles retain proper scales. Finally, the instructor should proofread the graphic with her fingers, not the eyes, before presenting it to the students. Be cautious if someone says that the graphic is "pretty" or "beautiful." This is not significant to the visually impaired student. The teacher should take another look and make sure that the graphics are accessible and clear.

## CHALLENGES AND FEASIBILITY OF INTEGRATING A BLIND STUDENT INTO PHYSICS CLASSES

Camargo and Nardi [1] describe four teaching activities in an educational environment attended by students with and without visual impairment.

The authors emphasize the difficulties/feasibility of the visually disabled student who was born blind. The difficulties/viabilities of students who were blind from birth are different from those of students who lost their vision later in life. This fact requires particularized analysis of the contexts where the loss of vision occurred.

The application of the teaching modules took place in two stages, the preparatory moment and the moment of defining the educational environment. The preparatory moment was characterized by planning the activities and the teaching material.

## CONCLUSIONS

The inclusion of visually impaired (VI) students in classrooms has led to the need to rethink teaching practices so that this inclusion can occur. When receiving VI students in physics and mathematics classes, teachers were faced with the novelty of learning to teach without being able to use visual resources. Visually impaired students benefit from education with a strong emphasis on language [10, 11]. Despite having the same learning abilities and curiosity as their sighted peers, they are often placed in lower levels of education compared to peers with similar capabilities. It is understood that teaching physics in fully inclusive classrooms requires that teachers have knowledge of appropriate materials to develop their own strategies with VI students. For VI students to achieve better living conditions, teaching and assessment strategies and instruments must be adapted to their reality.

## BIBLIOGRAPHY

1. de Camargo, E. P., & Nardi, R. (2009). A condução de atividades de mecânica para alunos com e sem deficiência visual: dificuldades e viabilidades/Difficulties and viabilities of having a born blind student effectively attend mechanics classes. *Acta Scientiae, 11*(2), 101–118.
2. de Camargo, E. P. (2016) *Inclusão e necessidade especial: compreendendo identidade e diferença por meio do ensino de física e a deficiência visual.* São Paul: Editora Livraria da Física.
3. Santos, O. D. A. M. (2020). Sequência Didática Sobre as Leis de Newton Para Alunos com Deficiência Visual, Master Dissertation, UFRJ/IF, Rio de Janeiro.
4. Software QuickTac™. https://duxburysystems.com/products.asp
5. Software Monet. https://www.acessibilidadebrasil.org.br/joomla/softwares?id=685
6. PROJETO DOSVOX. https://intervox.nce.ufrj.br/dosvox/

7. Tato, A. L. (2009). Material de equacionamento tátil para usuários do sistema Braille. Master Dissertation – Centro Federal de Educação Tecnológica Celso Sukow da Fonseca, Rio de Janeiro.
8. Azevedo, A. C. (2010). Construção Gráfica para Alunos Cegos. *A Física na Escola*, 11(1), 48.
9. Spence, D., & Osterhaus, S. A. (1997). *American Foundation for the Blind Braille Literacy Mentors in Training: The Next Generation – Teaching Special Codes: Nemeth, CBC, and Tactile Graphics.* Workshop in Fremont, California (August 7–9, 1997) and Atlanta, Georgia (September 11–13, 1997).
10. Sevilla, J., Ortega, J., Blanco, F., & Sánchez, C. (1991). Physics for blind students: a lecture on equilibrium. *Physics Education, 26*, 227. https://doi.org/10.1088/0031-9120/26/4/004
11. Toenders, Frank G. C., de Putter-Smits, Lesley G. A., Sanders, Wendy T. M., & den Brok, Perry (2017). Analysing the physics learning environment of visually impaired students in high schools. *Physics Education, 52*, 045027. https://doi.org/10.1088/1361-6552/aa737c
12. Castro-Castillo, D. C., & Tuay-Sigua, R. N. (2023). Revisión sobre la enseñanza de la física en estudiantes con diversidad funcional visual. *Revista científica, 46*, 1–16.
13. Lisboa, A., & Peña, F. J. (2023). Teaching labs for blind students: equipment to measure the inertia of simple objects. *arXiv preprint arXiv:2301.07033*
14. França, S., & Siqueira, M. (2019). Propostas didáticas no ensino de física para deficientes visuais: análise de trabalhos em periódicos e eventos nacionais (2000–2018). *Latin-American Journal of Physics Education, 13*(4), e001.
15. Goncalves, A. M. B., Cena, C. R., Alves, D. C. B., Errobidart, N. C. G., Jardim, M. I. A., & Queiros, W. P. (2017). Simple pendulum for blind students. *Physics Education, 52*(5), 053002.
16. Bulbul, M. Ş., Garip, B., & Ozdemir, Ö. F. (2017). Using a force concept inventory test with visually impaired and blind students. *European Journal of Physics Education, 6*(3), 20–31.
17. Reynaga-Peña, C. G., & del Carmen López-Suero, C. (2020). Strategies and technology aids for teaching science to blind and visually impaired students. *User-Centered Software Development for the Blind and Visually Impaired: Emerging Research and Opportunities.* IGI Global, 26–37.
18. Dickman, A. G., Martins, A. O., Ferreira, A. C., & Andrade, L. M. (2014). Adapting diagrams from physics textbooks: a way to improve the autonomy of blind students. *Physics Education, 49*(5), 526.
19. Abu Talib, R. I., Nikolic, P. K., Sunar, M. S., & Prada, R. (2020). In-visible island: inclusive storytelling platform for visually impaired children. *Mobile Networks and Applications, 25*, 913–924.
20. Bülbül, M. Ş., Eryurt, K., Mhaiskar, V., & Gupta, A. (2013). In touch with distance and displacement. *Physics Education, 48*(3), 275.
21. Bulbul, M. S., Garip, B., & Özdemir, Ö. F. (2015). Using a force concept inventory test with visually impaired and blind students. *European Journal of Physics Education, 6*(3), 20–31.

22. Hestenes, D., Wells, M., & Swackhamer, G. (1992). Force concept inventory. *The Physics Teacher, 30*(3), 141–158.
23. Henderson, C. (2002). Common concerns about the force concept inventory. *The Physics Teacher, 40*(9), 542–547.
24. Huffman, D., & Heller, P. (1995). What does the force concept inventory actually measure?. *Physics Teacher, 33*(3), 138–143.
25. Savinainen, A., & Scott, P. (2002). The force concept inventory: a tool for monitoring student learning. *Physics Education, 37*(1), 45.
26. Holt, M., Gillen, D., Nandlall, S. D., Setter, K., Thorman, P., Kane, S. A., … & Supalo, C. (2019). Making physics courses accessible for blind students: Strategies for course administration, class meetings, and course materials. *The Physics Teacher, 57*(2), 94–98.

# Appendix A

Some useful software to teach visually impaired students

## MATHML

MathML (Mathematical Markup Language) is an XML-based markup language designed to represent mathematical notation and structure. It allows math content to be displayed and interpreted on web pages and in other digital formats in a way that both humans and machines can understand. MathML describes both the presentation and the meaning (semantics) of math expressions. It can be rendered in web browsers (especially when combined with tools like MathJax), facilitates accessibility (e.g., for screen readers), enables math-aware search and indexing, and supports integration with other XML vocabularies, like HTML or SVG.

Example:

Here's a simple MathML snippet that represents the quadratic formula:

```
<math xmlns="http://www.w3.org/1998/Math/MathML">
 <mi>x</mi>
 <mo>=</mo>
 <mfrac>
   <mrow>
    <mo>-</mo><mi>b</mi>
    <mo>&#xB1;</mo>
    <msqrt>
      <msup><mi>b</mi><mn>2</mn></msup>
      <mo>-</mo>
      <mn>4</mn><mi>a</mi><mi>c</mi>
    </msqrt>
   </mrow>
```

```
    <mrow><mn>2</mn><mi>a</mi></mrow>
  </mfrac>
</math>
```

This represents:

$$x = \left(-b \pm \sqrt{\left(b^2 - 4ac\right)}\right) / 2a$$

## LATEX WITH MATHJAX

LaTeX with MathJax is a powerful way to display mathematical notation on the web using LaTeX syntax, made possible by the MathJax JavaScript library. LaTeX is well known typesetting system widely used in mathematics, science, and academia. It uses plain text and special commands to write math (e.g., x = \frac{–b \pm \sqrt{b^2–4ac}}{2a}). LaTeX is popular because it is concise, readable (to those familiar with it), and capable of expressing complex math.

MathJax is a JavaScript library that renders LaTeX (and MathML) on web pages, displaying beautiful, scalable math notation. It works in most modern browsers without requiring any plugins. It can also render MathML and AsciiMath, but LaTeX is the most commonly used format with it.

Example

If one write this in a webpage using MathJax:

```
\(x = \frac{–b \pm \sqrt{b^2–4ac}}{2a} \)
```

MathJax will render it like this:

$$x = \left(-b \pm \sqrt{\left(b^2 - 4ac\right)}\right) / 2a$$

## BRAILLE MATH CODE (NEMETH CODE)

Braille Math Code, more specifically known as the Nemeth Code, is a braille system designed to represent mathematical and scientific notation for people who are blind or visually impaired. It is Essential for STEM education for blind students. It extends the six-dot braille system to include symbols for arithmetic, algebra, geometry, calculus, and more. Nemeth Code uses special braille symbols to represent math operations, numbers, and structures (like fractions, exponents, square roots, etc.). Math expressions are written linearly, meaning they're read left to right, which is different from the way math might be spatially arranged on a printed page.

For example, a simple equation like:

$$x = \left(-b \pm \sqrt{\left(b^2 - 4ac\right)}\right) / 2a$$

It would be represented using a sequence of braille characters, with switching codes that tell the braille reader they are now reading math (not regular text), and when the math ends.

## MATHTALK

MathTalk can refer to different tools or concepts depending on the context, but generally, it's associated with making math more accessible, especially through speech and audio technology. The MathTalk program is an assistive technology that allows users to speak math out loud, and the system will translate it into mathematical notation. It's especially helpful for: students who are blind or visually impaired; people with physical disabilities who can't use a keyboard/mouse easily; and voice-controlled math input, for creating documents or solving equations. It converts spoken math into visual or braille output, works with speech recognition software like Dragon NaturallySpeaking, and supports Nemeth Code, MathML, and print math.

It is often used with screen readers and braille displays.

Example: You say

"x equals the fraction with numerator negative b plus or minus the square root of b squared minus four a c, and denominator two a"

And MathTalk converts it into:

x = \frac{–b \pm \sqrt{b^2–4ac}}{2a}

Alternatively, it can also convert it into Nemeth braille, or a format for screen reading.

## DRAGON NATURALLYSPEAKING (WITH MATH ADD-ONS)

Dragon NaturallySpeaking (often just called Dragon) is a speech recognition software developed by Nuance Communications. It allows users to control their computer and dictate text using their voice. When you refer to "NaturallySpeaking with math add-ons," you're likely talking about integrating math functionality with Dragon for dictating mathematical expressions or controlling math software via voice. Dragon NaturallySpeaking is used for: voice-to-text dictation (fast and accurate), voice commands to control applications, open files, browse the web, etc. It is especially useful for people with disabilities, writers, researchers, and professionals who prefer voice input.

Since Dragon isn't built specifically for math, add-ons or third-party tools are used to expand its capabilities in math-related areas. These might include: MathTalk, a companion tool for Dragon designed for speaking math notation aloud. It allows one to dictate math equations, calculus expressions, and even graphs.

## DESMOS AND GEOGEBRA: BOTH OFFER SCREEN READER–FRIENDLY MATH ENVIRONMENTS

Desmos is a free, web-based graphing calculator and suite math tools designed to help students and teachers explore math visually and interactively. It's especially popular in middle and high school classrooms, and it's known for its user-friendly interface and powerful graphing capabilities. Desmos plots functions, parametric and polar graphs, inequalities, and more with real-time updates. You can adjust sliders to see how graphs change dynamically. Visual Math Exploration: Great for seeing the effects of parameters in equations, transformations, and intersections visually.

GeoGebra is a free, interactive mathematics software used for teaching and learning math from primary school up to university level. It combines geometry, algebra, statistics, and calculus in one easy-to-use package. Here's a quick breakdown of what makes it special: It creates constructions with points, vectors, lines, polygons, conic sections, and more. You can move objects and see how the rest of the construction changes accordingly. Every geometric object has an algebraic representation. So, when you move a point in the geometry window, the coordinates in the algebra window update automatically (and vice versa). Plot functions explore derivatives, integrals, and limits interactively. Spreadsheets and Statistics: Analyze data with spreadsheets and perform statistical operations or regressions. CAS (Computer Algebra System): For symbolic computation—solve equations, expand/factor expressions, etc. GeoGebra also creates and manipulates 3D mathematical objects.

## JAWS (JOB ACCESS WITH SPEECH)

JAWS (Job Access With Speech) is a screen reader, a type of assistive technology designed for people who are blind or visually impaired. It allows users to interact with a computer by converting on-screen text into speech or Braille. JAWS reads aloud text from applications, web pages, documents, menus, and more. It works with Braille displays for tactile output. It also allows users to navigate and control a Windows computer using a keyboard and shortcut keys. JAWS supports web browsing, email, document editing, and other common tasks.

CHAPTER 5

# Teaching waves

## INTRODUCTION

Waves represent one of the foundational models in classical physics, with wave phenomena playing a significant role across various scientific and engineering disciplines. Concepts introduced during the study of waves often lay the groundwork for understanding quantum mechanics.

Research has shown that students frequently struggle with wave motion. Many adopt a model in which wave pulses behave like particles moving through a medium. These conceptual challenges are further exacerbated by the fact that waves are often students' first encounter with functions involving two variables. While this introduces mathematical complexity, the greater obstacle for most students is visualizing what waves look like and how they propagate through space. Static textbook illustrations, typically limited to one-dimensional transverse waves, make it difficult for students to imagine how a wave changes over time or across different positions. Visualizing longitudinal waves or waves in two or three dimensions proves even more challenging.

Classroom demonstrations of traveling waves are limited, as most setups naturally produce standing waves due to reflection at boundaries. For sighted students, tools like computer simulations and videos can be valuable for understanding wave motion.

DOI: 10.1201/9781003510079-5

Wavefronts and wavefront diagrams present another challenge. Since these diagrams typically depict a single moment in time, students may mistakenly associate wavefronts with a graph of the wave itself.

Wittmann [1] provides a detailed analysis of student reasoning in wave physics within a university classroom. He notes that students often misapply object-based reasoning to wave phenomena, relying on familiar concepts that are not appropriate for wave contexts. This tendency is reinforced by everyday experiences with finite-length waves, such as ocean waves. Research indicates that students focus on the finite size of waves, influencing their interpretations of wave superposition, propagation, and reflection. Rather than emphasizing the event-like nature of wave phenomena, students tend to adopt an object-like perspective.

This chapter explores methods for teaching wave phenomena to visually impaired students, offering detailed guidance on experiments and instructional strategies tailored to their needs.

## INCLUSIVE TEACHING OF WAVES IN HIGH SCHOOL

The models available for teaching waves, especially in high school, do not consider the actual stage of development of most students [2]. In addition, the difficulty in presenting frequencies, whether electromagnetic or mechanical (sound) in nature, that are different from the ranges that we can interpret through our senses, tends to constitute a huge epistemological obstacle. To shed some light on this problem, we developed two experiments that present relationships between invisible and inaudible frequencies, but which, through processing with the Arduino platform, can be converted into frequencies that sensitize our senses. In this way, we intend to reveal this invisible world, bringing experiments into the classroom that can even be used by students with visual or hearing impairments.

One difficulty I have noticed over the years of teaching physics is being able to explain to students that, although we cannot capture information through our senses, this does not mean that the information is not being sent. Electromagnetic radiation in regions with frequencies equal to or greater than ultraviolet, and equal to or less than infrared, cannot, as a rule, be captured by our eyes. Their effects and applications, however, are widely publicized. So, how can we acquire knowledge about what we cannot see or hear? The answer may seem trivial to those who are already initiated in the physical sciences, but it is not for most students.

Therefore, when teaching students, especially high school students, we must take their capacity for abstraction into account. We must have the necessary sensitivity to understand that physical models that deal with something that is beyond our sensations can be too abstract for their understanding, constituting a huge obstacle in the teaching and learning process [3].

Therefore, we must equip ourselves with different teaching methods so that, by applying them, we can make students approach scientific models, especially those that require greater abstraction, with a little more naturalness. The pedagogical proposal of developing activities based on the development of experiments, in which the aspect of social interaction is present, even if it does not occur in a laboratory, is defended in numerous works by teaching researchers such as Dewey et al. [4].

Piaget identified four main causes for the development of cognitive reasoning skills [5]. They are: maturation, experience with the physical environment, social transmission and balance. Maturation means that students must be biologically mature and physically developed. Experience with the physical environment involves previous concrete experiences, as well as the ability to remember these experiences. Piaget describes two types of experience, physical experience and logical-mathematical experience. The first is that obtained directly from objects, while the second occurs through actions that affect objects. In the social transmission stage, students must be able to communicate information in oral and written forms. In the equilibrium stage, an internal mental process occurs where new experiences are combined with previous expectations and give rise to new logical operations, that is, due to cognitive growth, students must be exposed to cognitively challenging situations, where their mental operating conditions are not adequate. The accommodation process, that is, the restructuring of assimilation schemes, also called equilibrium, is the process by which students deal with this new information resulting in cognitive growth. Accommodation is represented by newly acquired knowledge.

Research shows that students learn best through interactive involvement. In this article, we follow the Karplus learning cycle [6, 7]. The Karplus cycle transposes Piaget's theory into a model applicable to the learning process, incorporating the four factors mentioned in the previous paragraph. The Karplus learning cycle includes three instructional stages that combine experience with social transmission and encourage self-regulation. These three phases are Exploration, Introduction of the Concept, and Application of the Concept.

Stage 1, also known as Exploration, consists of a brief familiarization with the topic. Students familiarize themselves with equipment and think qualitatively about ideas that will be discussed in subsequent stages. In stage 2, or Concept Introduction, the instructor teaches the subject in the usual way, including concepts, derivations, and terminology, at a pace that students can assimilate and apply. Students are guided to discover principles that will later be discussed in stage 3. Finally, in stage 3, known as Concept Application, students make quantitative measurements to verify and apply what they have learned in the previous stages. The stages may also involve group problem-solving.

It is essential, however, that the teacher seeks to maintain student engagement from the beginning to the end of the activity, challenging them to find, by manipulating the experiment and using the corresponding physical models, answers to the questions proposed by the teacher at the beginning of the activity. Taking the wave model of sound and light as a reference, a possible guiding question for the class could be, for example: knowing that man has a limit (upper and lower in terms of frequency), both in terms of hearing and vision, would it be possible to hear and see beyond these ranges?

Of course, the answer would be no without the use of a transducer, but the goal is to challenge the student to reflect on the question. To do this, it will be necessary to understand the wave models of sound and light to produce a scientifically plausible answer. It will also allow a moment of reflection on the difficulties encountered by students with disabilities, both blind and deaf. In a mixed class, the student with a disability (blind or deaf) will be able to participate in the activities using other senses, so as not to be excluded from the activity, being guided by the teacher and receiving help from classmates.

We also observed that, according to Ferreira [6], experimental demonstration activities allow students to develop their observation and reflection skills, as well as helping them understand the content studied. In our work, they were developed with a focus on a qualitative approach, in which students, together and under the guidance of the teacher, will construct and interact with the experiments.

The purpose of the experiments developed here is to provide students with the means to experience frequency bands that our senses cannot detect, both inaudible and invisible. In this way, they will be able to accept more naturally that there is a world that is undetectable to our senses, and that, for it to be perceived, it is necessary to develop physical models that relate, using analogies, physical phenomena in our daily lives, such as waves in water that are related to electromagnetic waves. Evolutionary characteristics will also be addressed, such as the fact that, in nature, some animals had their senses developed differently from those of human beings and that, therefore, they have vision and hearing limits, in some cases, beyond our perception.

In the following two sections, we will work with the Arduino platform. An easily accessible prototyping platform that allows the user, through an integrated drive electronics and a board with input and output pinouts, to develop data acquisition, control and processing systems. This information is processed by an Atmel AVR processor. After processing, its output can be used to transmit the processing result [8].

Connected to the computer via a USB port, the platform can be programmed using the Processing language, where programs or, in Arduino language, "sketches" are typed [9–11]. The computer communicates with the Arduino board, producing both input and outputs of data. It is a simple but highly functional platform that can assist in various laboratory tasks, serving the study of various fields of science, including robotics.

For teaching physics, Arduino allows the implementation of various types of experiments at a relatively low cost, since it can be reused and reprogrammed as needed.

## USING THE KARPLUS CYCLE IN WAVE TEACHING

In this project, two applications related to the study of waves, associated with vision and hearing, were developed. The first aims to reveal a world of sound that is "invisible" to our ears, where we will use an ultrasonic sensor to illustrate the limitations of human auditory perception. In the second, we will have a barrier sensor using infrared light, therefore invisible to the human eye, which will produce, when activated, a noise and a light signal that will sensitize our hearing and vision, indicating the interruption of this beam. Both experiments allow for different didactic approaches, and can be used together or separately, including helping students with visual or hearing impairments learn physics.

To include students with visual impairments in the study of waves, some initial forms of approach were prepared that can contribute to the implementation of the concept, based on the use of low-cost resources. These activities can be prepared and put into practice in a mixed class, composed of students with and without disabilities, allowing for full integration of these students.

Initially, in the Exploration stage, a brief familiarization with the concepts is promoted. It is also an opportunity for students to familiarize themselves with the equipment and to think qualitatively about ideas that will be discussed in the following phases. In this first phase, the study of waves should be guided, initially, by the presentation of mechanical waves generally produced in strings or springs. This first contact with mechanical wave movement, which occurs before formal teaching, can bring about a positive affective experience when reproduced in the classroom. The benefit of the affective relationship in the school environment is very significant, being the object of study of several research in education. The depth of the affective relationships can be observed in their conceptualization by Dantas: "affectivity designates [...] the psychic processes that accompany the organic manifestations of emotion. Affection can well be conceptualized as one of the forms of love" [12].

In the Introduction to the Concept stage, the teacher presents the subject in the usual way, using the board, textbook, oral presentation, and/or small experiments that can be developed in a fun way with the participation of the entire class. The initial playful studies will build a bridge to later deepen the understanding of the wave phenomenon and its characteristics, such as period, frequency, speed, and wavelength. The first experiment that can be done with students is the generation of waves in a mechanical medium. In the case of blind students, they will first be generating their waves on a spring (it can be a toy spring, popularly known as a "crazy spring") and then they can gently slide their two hands along the spring's undulations, generated by another student, withdrawing their hand at the slightest contact, thus perceiving its geometric structure and propagation form.

Initially, a wave of length $\lambda/2$ can be produced, where the blind student can perceive its outline with both hands, starting from the ends, where we find the bellies (practically motionless) going toward the center, where he will notice that the hands will be increasingly upward, reaching the maximum point when both hands touch in the region called the wave crest. The same can be done for a wave of length $\lambda$, where the student will notice three nodes and two crests. It is important, during the experiment, to try to keep the frequency and amplitude of the wave constant, which will tend to lose energy when touched, even if very gently. Through touch, these students will be able to perceive the rise and fall of the spring and build a mental model of this wave form encompassing the way it propagates, starting from the disturbing agent toward the opposite end. This analysis can be reinforced through a two-dimensional model of a sinusoidal wave of length $\lambda$ (a string or shoelace in the shape of a sinusoidal wave can be glued to a piece of cardboard), for example, as can be seen in Figure 5.1.

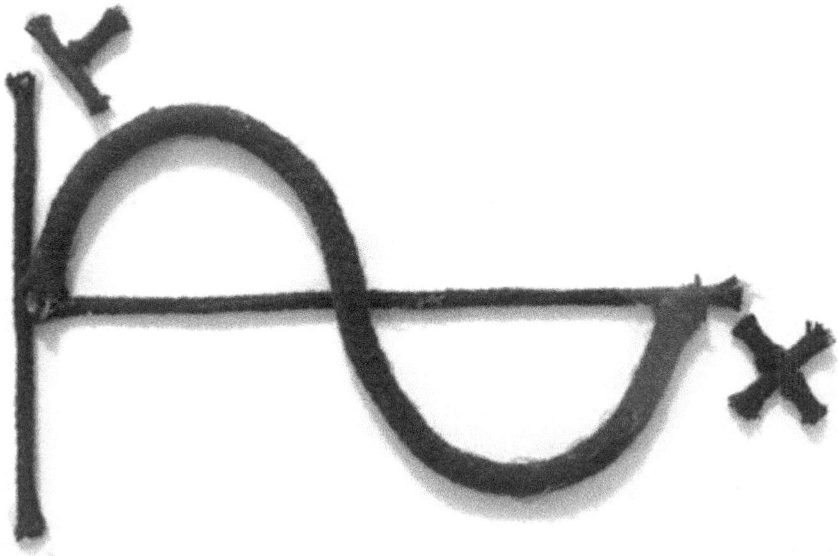

FIGURE 5.1 Two-dimensional model of a wave of length λ, made with thick round lace glued onto cardboard. Axes are made with thin lace. Reproduced with permission from [2].

This experience will also be of great value to sighted students, who will be able to follow and participate in the production of these waves and better understand their characteristics.

Manipulating the spring and string can contribute to the construction of other concepts such as valley, crest, node, amplitude, and wavelength. Another characteristic of this wave movement refers to the concept of a transverse wave, where the disturbance moves perpendicular to the direction of propagation. For other students, these wave characteristics can be drawn in the traditional way on the board.

The spring can also contribute to the understanding of the longitudinal wave model, in which the deformation of the spring moves in the same direction as the propagation. For blind students, the spring can be stretched while keeping a small portion of it deformed (compressed), tied with string or tape. Using touch, the student will be able to perceive this deformation, which differs from the rest of the spring due to the proximity of the links, as being the disturbance that travels through the spring. This disturbance travels along the spring in the same direction as its propagation, and is therefore an example of a longitudinal wave, unlike the wave we saw previously, whose disturbance is transverse to the direction

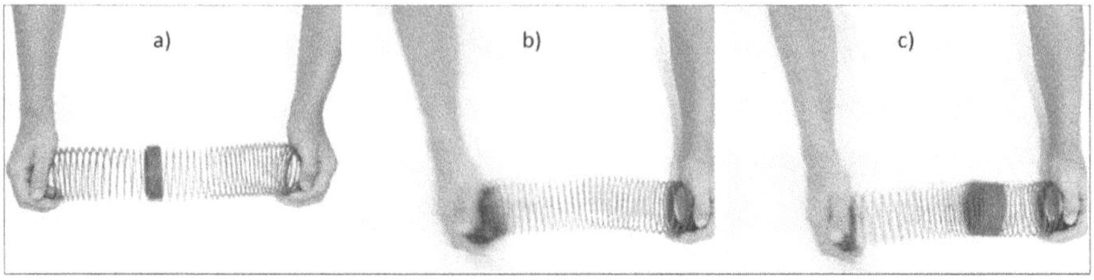


FIGURE 5.2 (a) Image of the spring kept compressed in the central region by adhesive tape. (b and c) A pulse traveling through the spring. Reproduced with permission from [2].

of propagation, as can be seen in Figure 5.2a. By supporting the partially extended spring on a flat bench and holding it by both ends, the blind student will be able to produce a disturbance at one of the ends and perceive, through the sound produced, the movement of this pulse, traveling the entire length of the spring. For the effect to be perceived clearly, the environment must be silent. The other students will be able to visually perceive the phenomenon, which is therefore an interesting way to illustrate the behavior of a longitudinal wave to the entire class. Figure 5.2b and c illustrates this experiment.

After this macroscopic introduction to the longitudinal wave model, we can delve into the microscopic model of the vibration of air molecules, as a mechanical support for the propagation of sound. At this point, it may be interesting to comment on the vocalization of sound through the vibration of the vocal cords and the capture of this sound by the auditory system until the information is sent to the brain, its limiting ranges for capturing sound frequencies and the comparison with the frequency ranges perceived by some animals. The purpose of this comparison is to understand, at the end of the class, that our senses are limited, in the case of hearing, by a frequency range that varies from 20 to 20,000 Hz and that to detect sounds above or below this range we need specific detectors. Therefore, the fact that we are not hearing a noise does not mean that this noise is not being propagated.

After the theoretical introduction to the wave model, the experiment "The sound that the deaf can see and the blind can hear" can be introduced, initially presenting the components that make up the experimental setup. The main component is the ultrasonic sensor, which is responsible for producing a sound that is inaudible to us and, at the same time, is equipped with a sensor capable of detecting the return of this produced sound, when it hits an obstacle and returns to it. The explanation of how it works refers to the concept of echo, in which the reflection of a wave and the speed of propagation of sound in the air, which varies according to its temperature, can be explored. Observing the sketch, to calculate the distance, we need to inform the speed of sound in the environment where the sensor is being used. This is done through the command "distance = (duration/2) * 0.0343," which will be discussed later.

Finally, in the Concept Application stage, students will answer a questionnaire, which should be prepared by the teacher, to verify and apply what they learned in the previous stages. Some questions may be of a qualitative (conceptual) nature, with the main objective of assessing whether the student can distinguish characteristics of the two types of wave models discussed (mechanical and electromagnetic). And quantitative questions, which may involve calculations of frequency, period and wavelength. In addition to the propagation speed, which can also be used to illustrate exercises using the sonar concept.

## EXPERIMENT 1—THE SOUND THAT THE DEAF CAN SEE AND THE BLIND CAN HEAR

Sound and light share some oscillatory characteristics, such as frequency, period, and length. However, when the similarities end, we must reinforce to students the differences that consider the form and the propagation medium so that we can use the analogy within its field of limitation, with a perfect delimitation of its boundaries. For light, as an electromagnetic wave, there is no need for a physical medium for its propagation and the wave disturbance occurs transversally. In the case of the sound wave, in addition to propagating longitudinally, there must be a medium to support its propagation, without which the sound ceases to propagate. At this point, the first experience, the one that young people experience when watching science fiction films in which explosions in space generate enormous bangs, can become a huge epistemological obstacle, that is, these first contacts with pseudoscientific experiences, when not subject to criticism, end up becoming truths that are difficult to purge, creating psychological conditions that hinder the progress of the scientific spirit [4]. The professor, with his experience, must be emphatic in stating that the characteristics of sound do not allow it to propagate where there is no material medium to support it, unlike light, given that we are bathed in rays of light that cross the immensity of space before reaching us. However, even outer space, which is constantly used as an example of absolute vacuum, contains matter through which sound can propagate. In some regions of space, such as in an interstellar cloud, there may be a density of about ten hydrogen atoms per $cm^3$, which allows the propagation of sound waves with wavelengths comparable to the dimensions of the solar system [12]. These waves would not be detected by our ears, however, once again, it is necessary to clarify the important concepts in this matter, which are the definitions of absolute vacuum and the respective limits of our hearing.

Our hearing allows us to capture and hear sounds in the frequency range of 20 Hz to 20 kHz. Frequencies below 20 Hz are called infrasound and frequencies above 20 kHz are called ultrasounds. The closer to 20 Hz, the lower the sound is perceived by us; conversely, the closer to 20 kHz, the higher the sound becomes. Other animals can capture sounds in different frequency ranges, such as elephants and bats. Infrasonic frequencies, below 20 Hz, have little scattering, which facilitates their propagation, allowing some animals to communicate over long distances. Ultrasonic frequencies, on the other hand, are strongly scattered by small objects and, therefore, are widely used in medicine to generate images of the inside of our bodies. With them, it is possible to measure the dimensions and shapes of various organs without the need for surgical interventions and without the possibility of cellular damage, as occurs with X-rays.

Returning to the animal world, bats can capture sounds from 10 to 120 kHz, so when they produce sounds of around 50 kHz, they can capture their echo and, in this way, have a perception of the spatial location of objects in front of them, like a sonar in a submarine. It is as if they could see with their ears in an environment devoid of light, such as in the darkness of a cave [13].

## THE EXPERIMENT

The following experiment aims to introduce a device into the classroom in which students can be in front of a sensor that produces a sound that is inaudible to humans and that can capture its echo, generating information. This information is generated through colored lights that identify the distance to an obstacle by simply alternating the lighting of LEDs of different colors and the production of noise. In this way, we will be showing students that we need the right sensors to capture information and, likewise, to be able to interpret it. Using an electronic component called a buzzer, an audible sound of different tones can be generated for each distance, which would favor teaching blind students. In the case of other students, lighting the colored LEDs would also allow a comparative approach between vision and hearing, showing students the differences and similarities between the wave models of light and sound. In a way, this experiment could, in an inclusive way, put deaf and non-deaf students on an equal footing to understand the phenomenon, since the sound frequencies are inaudible to both. The visual stimulus would also benefit the deaf student.

The ultrasonic sensor used, HC-SR04, emits noise in the 40 kHz range and operates with a voltage of 5 VDC, being able to detect barriers at a distance ranging from 2 to 400 cm with a resolution of 0.3 cm. It operates at an effective angle of less than 15°. More details can be found in reference [14].

The device has two cylindrical drums, one being the emitter and the other the receiver, which correspond to a speaker and a microphone, respectively. When programmed and powered, the emitting component sends sound pulses into the environment. Like a sonar, when this pulse encounters an obstacle, within the operating limits, it is reflected to the sensor. The receiving component is able to capture it, and the system calculates the time elapsed between the emission and reception of this signal. Knowing the time between emission and reflection and the speed of propagation of sound in the medium, in this case air, the distance at which the obstacle is from the sensor can be determined. An illustration of how it works can be seen in Figure 5.3.

In addition to the ultrasonic sensor, three colored LEDs are needed. Green, red, and blue LEDs, three 330 Ω resistors, and several wires adapted for connection to the protoboard and the Arduino board were used.

Another component used is the buzzer, which is an electronic device capable of producing sound when subjected to a potential difference (ddp), that is, a transducer that can convert electricity into mechanical movements capable of generating sound. It is widely used in electronic

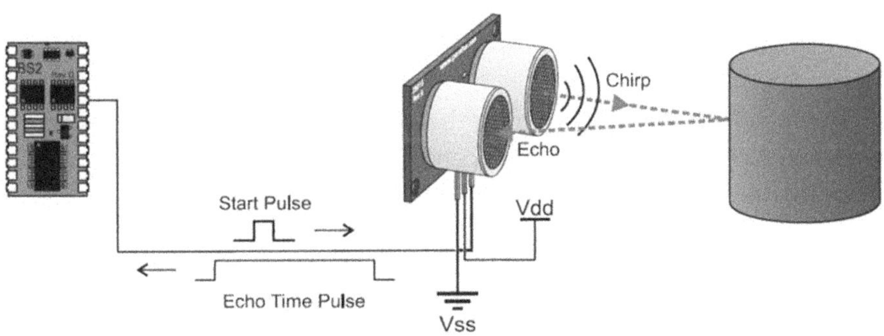


FIGURE 5.3 The sensor emits a noise that, when reflected off an obstacle, returns to the sensor, being captured by it [15]. Reproduced with permission from [2].

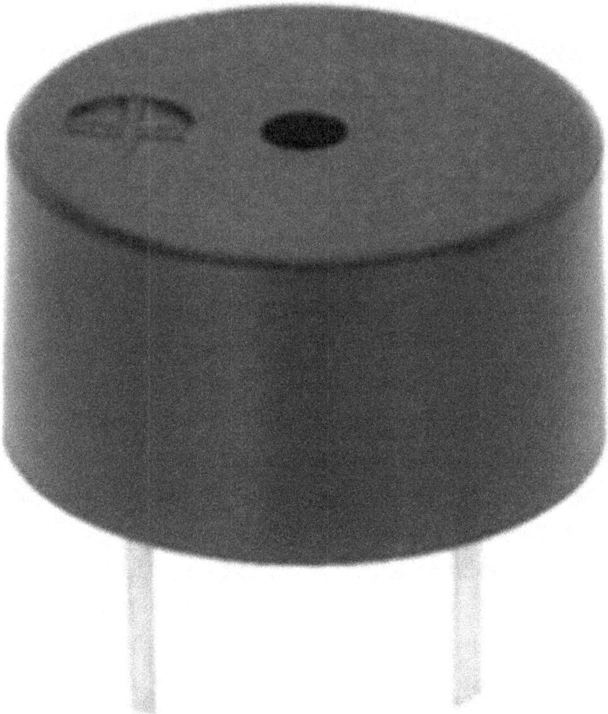

FIGURE 5.4 Encapsulated buzzer [16]. Reproduced with permission from [2].

equipment and its purpose is to produce an audible noise that can be used as an alert, like a siren, or some type of indication. They can be of various types, such as piezoelectric or electromechanical [16].

Piezoelectric buzzers are easy to find at a relatively low cost. For this reason, they were adopted in the experiment circuit. Figure 5.4 shows the encapsulated buzzer used. The assembly with the connections of the components on the protoboard and the Arduino board can be seen in Figure 5.5.

The sketch, that is, the program that will be run on the Arduino, is presented in Appendix 1. No further comments are made about its command since the idea here is not to teach this programming language. However, we draw attention to the command {distance = (duration/2) * 0.0343}. This command expresses in computer language the knowledge that we extract from understanding the physical world. The speed of sound propagation varies according to the characteristics of the medium in which it is propagating. In air, this speed is also affected by temperature variations, with a speed of approximately 343 m/s being accepted for a temperature

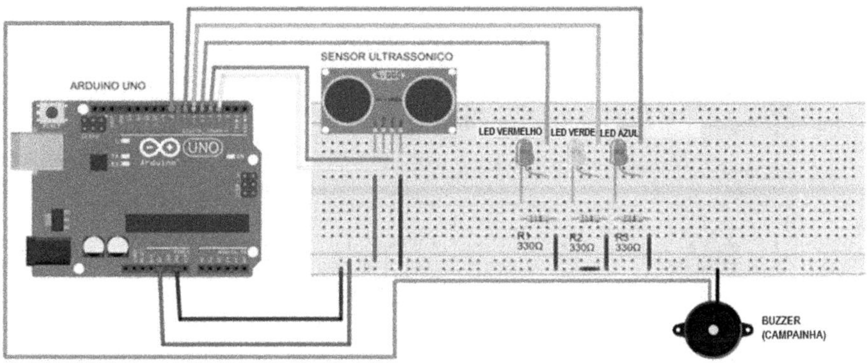


FIGURE 5.5 Assembling the components on the protoboard and Arduino board. Reproduced with permission from [2].

of around 20°C [17]. To know the distance at which the object is located, we will use this speed, together with the time count from the emission of the pulse until its return to the sensor. Since we want to know the distance (distance) at which the object is located and not the distance traveled by the pulse, we will use half the time (duration/2). We know that the distance can be calculated using the equation (d = t * v), that is, the distance traveled is equal to the product of the speed and the time spent in its movement. It is important to note that the Arduino platform works with pulses in the order of milliseconds and that we want the result of the distance in centimeters. To do so, it is necessary to divide the speed 343 m/s by 10,000 so that we have the result expressed in cm. For this reason, we see in the command half the time (duration/2) being multiplied by 0.0343.

The program activates the ultrasonic sensor so that it sends an audible signal from time to time. When the sensor captures the echo, the system automatically calculates the distance, and if this distance is greater than 30 cm, it activates the blue LED. When the distance is between 6 and 30 cm, it activates the green LED, and for distances less than 6 cm, it activates the red LED.

The working principle of each component of the experiment should be described to the student and, after the assembly is complete, using their own hands as an obstacle, they will be able to activate the LEDs and hear the sounds produced. The intention is for them to realize that a sound of a frequency above that which we can capture is in fact being emitted. And that, although we cannot hear it, the experiment proves that this sound is really being emitted. This fact is confirmed by the lighting of the LEDs and the sounds produced by the buzzer, which can be perceived by our senses.

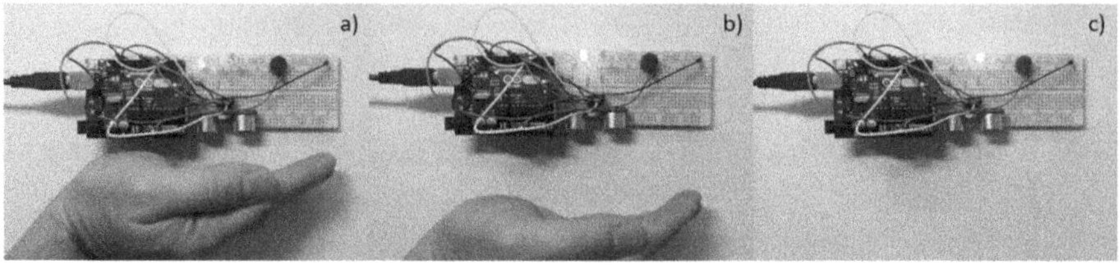


FIGURE 5.6 (a) With the palm of the hand at less than 6 cm, the red light is observed to turn on and a very low-pitched noise is emitted by the buzzer. (b) At a distance slightly greater than 6 cm and less than 30 cm, we can observe the green light to turn on and the noise becomes higher pitched. (c) By removing the hand from in front of the sensor and leaving it more than 30 cm away from any object in front of it, the blue light is observed to turn on and the buzzer produces a very high-pitched sound. Reproduced with permission from [2].

The experiment also provides an argument for working with students on calculations regarding the propagation and reflection of sound, contributing to reflective practice on the concepts covered.

In Figure 5.6, we see the experiment set up. In Figure 5.6a, with the hand at less than 6 cm, we observe the red light turning on and the others are off, the buzzer emits a low sound. In Figure 5.6b, at a distance greater than 6 cm and less than 30 cm, we observe that only the green light turns on and the noise becomes a little higher pitched. In Figure 5.6c, there is nothing closer than 30 cm to the sensor, this can be verified because only the blue light is on and by the sound emitted by the buzzer, which is extremely high-pitched.

When working with blind students, it is very important that during the explanation of how each component works and the assembly of the experiment, these students can, with the help of the teacher and their classmates, actively participate, handling and helping other classmates. In this way, with active participation, they will be able to better understand how each component works and feel like an integral part of the construction of this experiment so that, during its operation, greater interest and effort will be awakened in trying to use the concepts they learned during the study of waves. This active participation will also be important for using this experiment with deaf students. However, it is worth emphasizing that we are all deaf in relation to the sound emitted by the ultrasonic sensor and that the visual stimulus, through the colored LEDs, will be essential for the perception and understanding of the phenomenon, as well as the sound activation of the buzzer.

## EXPERIMENT 2—THE LIGHT THAT THE BLIND CAN HEAR AND THE DEAF CAN SEE

Just as our hearing is effective for a well-defined frequency range, the same occurs with vision. Human vision is associated with electromagnetic radiation in wavelengths ranging from 400 to 700 nm, making up what we call visible light [18]. In the animal kingdom, responses to electromagnetic stimuli can be quite different. Bees, for example, can see in the range from ultraviolet to yellow, and are therefore blind from red onward [19].

There are species of hummingbirds that can discern different shades of orange, imperceptibly different to the human eye, in bromeliad flowers that indicate a higher concentration of nectar. At the other extreme, we have animals that can see in the infrared region, such as some species of fish. We humans, who have three types of cones (or four, in the case of some women [20]), can only see in the visible region, except in surgeries to remove the lens, which increases sensitivity to a small range of ultraviolet. This was the case of the famous French painter Oscar-Claude Monet, who, having undergone cataract removal surgery, began to perceive this part of the spectrum [21].

## THE EXPERIMENT

In Figure 5.7a and b, we can see two types of infrared light-emitting diodes. The first has a slightly bluish encapsulation and the second is transparent. These diodes are widely used in remote controls for televisions and other electronic devices. In Figure 5.7c, we have an infrared phototransistor, which has a darkened body to prevent interference from frequencies other than the infrared range. Reception is possible because infrared can

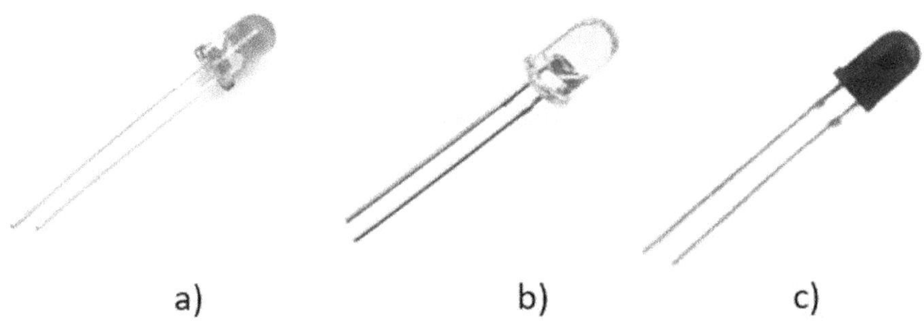


FIGURE 5.7 (a) Infrared emitting LED with a blue body. (b) Emitting LED with a transparent body. (c) Phototransistor. Its darkened body allows light to be filtered, allowing only infrared radiation to pass through [22]. Reproduced with permission from [2].

pass through the darkened encapsulation, as occurs in murky waters, and can be seen by some species of fish.

The phototransistor is an electronic device that has two terminals; one is the collector and the other is the emitter. The base is the part of the phototransistor that is exposed to light. When the base receives enough infrared light, it becomes active, allowing the passage of electric current from the collector to the emitter. Without sufficient light, there is no current conduction, thus leaving the emitter and collector electrically isolated.

Figure 5.8 shows the assembly diagram of the components on the protoboard and on the Arduino board. The circuit is basically an infrared barrier sensor that, when activated, produces an audible noise and causes a red LED to light up at the same time. The infrared emitting LED receives direct power from the Arduino board (5 VDC). To limit the current on the emitting LED, a 330 Ω resistor was used, as well as for the red LED. The buzzer has its negative terminal connected to the ground of the Arduino board and the positive terminal on pin 13, which will send +5 VDC when the barrier is interrupted. The detailed electronic schematic is shown in Figure 1 of Annex 2.

To form a barrier sensor, the infrared LED light must be pointed at the phototransistor so that it receives enough infrared radiation at its base, thus allowing current to pass from the collector to the emitter. The phototransistor is assembled according to the diagram in Figure 5.8. We can see that there is a 10 kΩ resistor connected to the collector of the phototransistor, while the emitter is connected to ground. A connection is made between the resistor and the collector to pin 2 of the Arduino board. When receiving light, the phototransistor allows the passage of electric

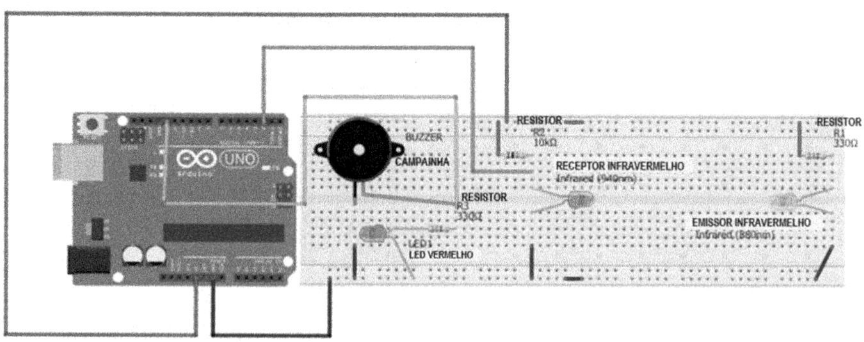


FIGURE 5.8 Assembly diagram of the experiment the light that the blind can hear and the deaf can see. Reproduced with permission from [2].

current, without any significant voltage reaching pin 2, since the collector and emitter, in this circumstance, are short-circuited, supplying the 10 kΩ resistor. The function of the 10 kΩ resistor is precisely to prevent the Arduino board from being short-circuited. Without receiving sufficient infrared light, the phototransistor does not conduct, sending the 5 VDC from the power supply to pin 2. When this occurs, that is, when voltage arrives at pin 2, Arduino interprets this information and directs +5 VDC to pin 13, activating the buzzer. It should be noted that in the infrared emitting LED, as in other LEDs, the longest terminal, which is the anode, must be connected to the positive pole of the power supply. However, in the case of the phototransistor, the longest terminal is the emitter, which must be directed to ground, as shown in Figure 5.9 [23]. The phototransistor is an electronic component widely used in educational experiments, both for the construction of spectrometers [24] and for barrier sensors that allow calculating the speed and its variation over time [25].

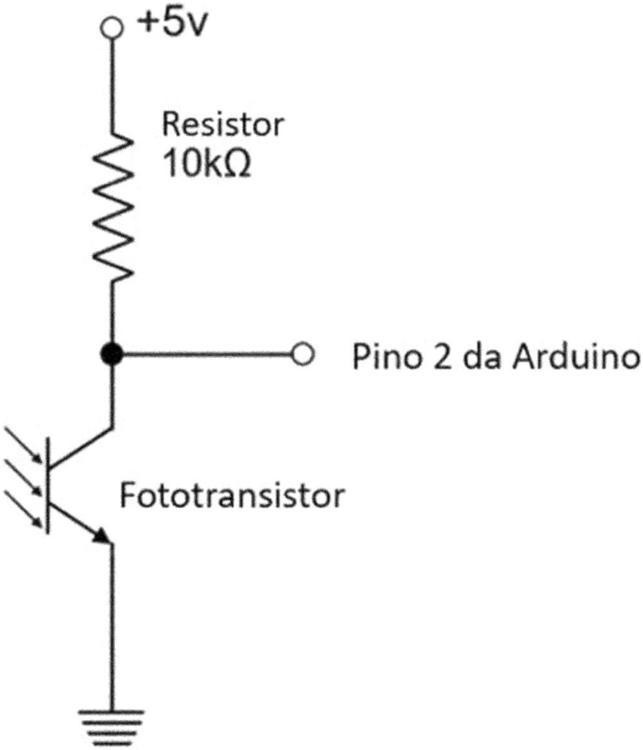


FIGURE 5.9 Circuit diagram with phototransistor. Reproduced with permission from [2].

The sketch for this experiment is very simple. With just a few commands, you can program the Arduino to perform the function. The sketch available in Appendix 2 can be copied exactly as it is reproduced and compiled for the Arduino board.

With the experiment assembled, as seen in Figure 5.10a and b, the infrared LED on the right illuminates the base of the phototransistor. This way, the circuit is ready and will remain so until the light beam is interrupted. The protoboard with the circuit can be placed inside a box so that students can observe that no visible light comes out of the infrared emitting LED.

To confirm that light is being emitted by the infrared LED, students can activate their cell phone cameras, as shown in Figure 5.10d, thus visualizing something that our eyes cannot see.

The audible warning and the red LED that are activated every time the light beam is interrupted allow students to reflect, once again, on the characteristics of light and sound. In the case of visually impaired students,

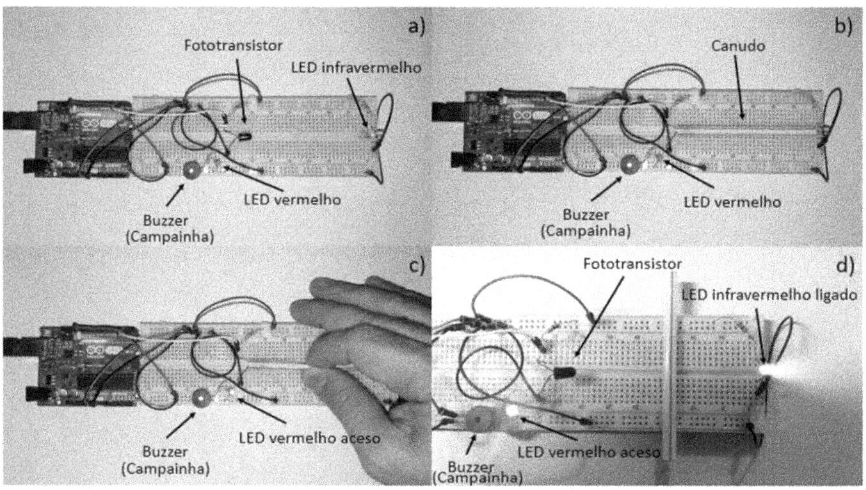


FIGURE 5.10 (a) All the components connected to the protoboard, as well as the connections to the Arduino board, with emphasis on the infrared LED and the phototransistor. Visually impaired students can feel both to see that one points directly at the other. (b) We inserted a straw between the infrared LED and the phototransistor. (c) By pressing the straw with your fingers, the phototransistor stops receiving the light from the infrared LED, thus activating the red LED and the buzzer. (d) Using a cell phone camera and pointing the infrared LED upward, it is possible to see that it remains on when the circuit is in operation. In this way, the phototransistor does not receive the light from this LED, activating the red LED and the buzzer. Reproduced with permission from [2].

it is necessary, once again, that the entire assembly be prepared with the participation of these students. The LEDs must be touched by blind students so that they perceive a certain directionality of the light emitted (or received, in the case of the phototransistor), through the lens located at its end. Due to this directionality of the light beam, it is necessary to arrange both components (infrared emitter and receiver) frontally, so that the phototransistor can receive enough infrared light from the infrared LED emitting it, which will allow it to function correctly. After the two components are positioned frontally, to make it easier to understand that both components are positioned frontally, a straw cut to the length of the distance that separates them (Figure 5.10b) can be used, fitting the ends into both components. In this way, the blind student will be able to perceive unequivocally, by touch, that the components are pointing toward each other. In addition, he will perceive, by the diameter of the straw (like the diameter of the LED) that the portion of the infrared beam that goes from the emitting LED to the phototransistor is contained inside this straw. Therefore, by slowly pressing the straw (with the thumb and index finger), the student will gradually be interrupting the transmission of this beam, until the moment when his fingers meet. At this moment, the buzzer will be activated, indicating that the beam has been completely interrupted. In the case of a deaf student, the red LED, activated together with the buzzer, will indicate the interruption of the beam.

As a result of using this work, the teacher can lead students to realize (in the case of the first experiment) that the ultrasonic sensor used to detect distances can illustrate, for example, how bats can move using their sense of hearing to "see" the world without light. This type of analogy can be used in the classroom and, as a complement, how sonar works can be explained. Just as one cannot see with one's eyes in the absence of light, it is very difficult to see in murky waters and at great depths. Therefore, boats use sonar, which is a device that works according to the same principle, that is, it sends a high-frequency sound signal in the direction it wants to measure and calculates the distance based on the time it takes for the echo to be captured.

A complementary activity can be worked on with the student, in which the fact that sound travels at a much higher speed in water (around 1,480 m/s) than in air (around 343 m/s) [17] is used, which allows for a faster calculation of the distance to an obstacle. On the other hand, compared to vision, since light travels in air at a speed of approximately 300,000,000 m/s, hearing produces a perception very slowly. For this reason, we see lightning (light) before hearing thunder (sound).

## DIDACTIC SEQUENCE TO TEACH PHYSICS OF WAVES TO BLIND STUDENTS

Lomas et al. [26] proposed using didactic sequences as a tool to bridge classroom learning with the lived experiences of blind students. The goal of the sequence was to help students connect theoretical concepts with real-world experiences, enabling them to perceive wave phenomena through multiple senses.

The sequence was divided into three stages, designed to span approximately two class hours in a typical basic education setting, though the duration could be adjusted based on the needs of both students and teachers.

In the first stage, a mental exercise called *Feeling the Waves* was introduced. This exercise involved students using their fingers to create a spatial representation of a transverse mechanical wave on a flat surface. Using two rulers or similar rectangular objects as barriers to define the area, the teacher guided the student's index finger in a perpendicular back-and-forth motion between the barriers. This demonstrated the boundaries and periodicity of the movement, drawing parallels to the behavior of mechanical waves and their propagation.

Following this, the teacher assisted the students in moving their hand parallel to the barriers, combining the motions to illustrate the shape and movement of a mechanical wave. Through this tactile activity, students were able to conceptualize and describe a mechanical wave using their own fingers. During this stage, key wave properties such as crests, troughs, frequency, and wavelength, along with their characteristics, were introduced and explored.

The second stage focuses on representing the phenomena of reflection and refraction through tactile-sensory stimuli. Three relief representations of these phenomena were created using wood, nails, and string. Different textures were incorporated into the materials to distinguish between the components of the models.

To help the student understand and interpret the representations, the teacher begins by guiding the student's hand over each element, explaining its role within the system. In the models, the strings connected by nails symbolize the path of light and its trajectory in space. An acrylic plate in the models serves as an analog for a glass surface or any medium where refraction occurs. For the reflection model, sandpaper represents the reflective surface. The angles are depicted using EVA foam, allowing students to feel the size differences between them.

The selection of materials was carefully made to prioritize the student's safety and comfort while handling the models. This tactile approach ensures an engaging and accessible way for visually impaired students to explore and understand these wave phenomena.

During the third and final stage of the didactic sequence, it is recommended that the teacher uses two acoustic tuning forks, sticks, and two amplifier boxes to demonstrate the phenomenon of resonance. When struck, the tuning forks will vibrate, and if they share the same frequency, the resonance phenomenon will occur.

To begin, the teacher should introduce the amplifier boxes, sticks, and tuning forks to the blind student, both individually and in their assembled form. With guidance, the student should create disturbances in the system by striking the top of one tuning fork with the stick. Through the resulting vibrations and auditory stimuli, the student will observe the resonance phenomenon.

It is important for the teacher to encourage the student to independently explore the materials while prompting them to describe their observations and experiences during the activity.

## ANOTHER PEDAGOGICAL ADAPTATIONS AND MATERIAL DESIGN FOR PHYSICS INSTRUCTION FOR VISUALLY IMPAIRED LEARNERS

In the absence of a braille textbook, it is possible to translate into braille, or to read and record some paragraphs of conventional physics textbooks, so that the student can read or listen to the content. However, it is important to revise and adapt sections in the textbook. One suggestion is to reduce the number of words, removing irrelevant content, such as irrelevant assignments and undoable experiments [27].

Toenders et al. [27] have demonstrated that the redesigning and supplementing braille study materials can significantly enhance the learning experience and accessibility of physics for visually impaired students in mainstream high schools. By removing non-essential content, adapting visual figures, and incorporating tactile models, valuable insights can be gained into how physics can be more effectively taught to students with visual impairments.

On the other hand, excessive adaptation may isolate the visually impaired student from the standard curriculum followed by their peers. Therefore, any redesign of materials must be iterative, as new limitations—such as the complexity of interpreting relief graphs—emerge during implementation.

Particular attention must be paid to assignments that reference visual phenomena, as these must be recontextualized using tactile or auditory equivalents. Collaboration with physics educators is essential in creating alternate examples that convey the same principles in ways that are accessible through touch or other senses.

In some cases, key visuals or texts must still be included in the braille version to allow peers or instructors to assist the visually impaired student in accurate interpretation. Without this support, even well-adapted materials risk reinforcing misconceptions.

Designing effective tactile is another critical aspect of accessible physics instruction. Proper orientation of these models is crucial, as incorrect placement may obscure braille labels or distort the intended meaning. Teacher or peer support is essential to ensure that students interact with these models as intended, and teachers must verify that interpretations are accurate to avoid conceptual misunderstandings.

Visually impaired students require more instructional time to grasp physics concepts than their sighted peers. Despite improvements in material design, their cognitive processing of tactile and auditory inputs, which are sequential rather than simultaneous, unlike sight and sound, demands a different pedagogical pace. The effort to adapt and improve study materials yields tangible benefits in understanding and engagement for visually impaired learners.

## FINAL CONSIDERATIONS

The lack of adequate teaching resources for students with visual impairments significantly hampers the teaching and learning process, often rendering it ineffective or entirely absent. Therefore, it is essential to prioritize the implementation of measures that genuinely promote the inclusion of these students within the school environment.

One of the greatest difficulties encountered by visually impaired physics students is related to conducting experiments. To address this problem, some experiments and didactic sequences were presented that allow the teacher to work on the wave concepts, relationships between invisible and inaudible frequencies that, through processing with the Arduino platform, can be converted into frequencies that sensitize our senses.

The proposed activities allow visually or hearing-impaired students to establish their relationships with learning physics through a continuous process that leads to imbalance—in the Piagetian sense—acquiring more appropriate concepts as well as more elaborate reasoning patterns.

The experiments help to illustrate the limits of our perceptions and some properties of sound and light, showing the existence of a universe beyond our senses. Through an inclusive experimental teaching proposal, it allows a sensory experience to blind and deaf students in mixed classes.

## BIBLIOGRAPHY

1. Wittmann, M. C. (2002). The object coordination class applied to wave pulses: analysing student reasoning in wave physics. *International Journal of Science Education, 24*(1), 97–118.
2. Silveira, M. V. D., Barthem, R. B., & Santos, A. C. D. (2018). Proposta didático experimental para o ensino inclusivo de ondas no ensino médio. *Revista Brasileira de Ensino de Física*, 41, e20180084.
3. Diniz, R. T. (2016). Usando Experimentação No Ensino Potencialmente Significativo De Óptica Geométrica. Dissertação de Mestrado, Universidade Federal Fluminense, Niterói, Rio de Janeiro.
4. Bachelard, G. (1996). *A formação do espírito científico: contribuição para uma psicanálise do conhecimento.* Rio de Janeiro: Contraponto.
5. Borges, A. T. (2002). Novos rumos para o laboratório escolar de ciências. *Caderno Brasileiro de Ensino de Física*, 19, 294.
6. Azevedo, A. C., & Santos, A. C. F. (2014). Learning cycles in physics teaching for visually impaired students. *Revista Brasileira de Ensino de Física*, 36, 4402.
7. Karplus, R. (1962). The science curriculum-one approach. *The Elementary School Journal, 62*(5), 243–252.
8. Ferreira, N. C. (1978). Proposta de laboratório para a escola brasileira: um ensaio sobre a instrumentalização no ensino médio de física. Dissertação de Mestrado, Universidade de São Paulo, São Paulo.
9. Margolis, M. (2011). *Arduino Cookbook*, 2nd ed. Sebastopol: O'Reilly Media.
10. Massimo, B. (2009). *Getting Started with Arduino.* Sebastopol: O'Reilly Media.
11. Dantas, H. (1990). *A infância da razão.* São Paulo: Editora Manole, 10.
12. Silva, S. T. (2011). Propagação do som: Conceitos e Experimentos. Dissertação de Mestrado, Universidade Federal do Rio de Janeiro, Rio de Janeiro.
13. Rodrigues, F. V., & Collino, R. (2008). Fisiologia da Audição. *Livro do V Curso de Inverno: Tópicos em Fisiologia Comparativa.* São Paulo: Universidade de São Paulo.
14. Fuentes, V. M. (2014). Introducción a la plataforma Arduino y al Sensor ultrasónico HC-SR04. *Profesional, Universidad Carlos III de Madrid.*
15. 38-3D. acesso em agosto de 2017. https://arduinosensors.com/index.php/arduino-ultrasonic-distance-sensor/
16. Das, S., Patro, S., Das, R., & Mishra, A. (2019, April). Arduino based safety device for the visually challenged. In *2019 International Conference on Communication and Signal Processing (ICCSP)* Chennai, India, (pp. 0601–0605). IEEE.

17. Freitas, M. A. B. (2005). Medindo a Velocidade do Som com o Microfone do PC, Final Course Project, Physics Institute, UFRJ, available at: https://www.if.ufrj.br/~carlos/trablicen/marco/monografiaMarco.pdf, Accessed from June 2025.
18. Jacobs, G. H. (2008). Primate color vision: a comparative perspective, *Visual Neuroscience, 25*, 619.
19. Ibarra, N. H., Vrobyev, M., & Menzel, R. (2014). Mechanisms, functions and ecology of colour vision in the honeybee, *Journal of Comparative Physiology. A, Neuroethology, Sensory, Neural, and Behavioral Physiology, 200*, 411.
20. Jameson, K. A., Highnote, S. M., & Wasserman, L. M. (2001). Richer color experience in observers with multiple photopigment opsin genes, *Psychonomic Bulletin & Review*, 8, 244.
21. Zimmer, C. (2012). *Monet's Ultraviolet Eye*, disponível em. https://www.downloadtheuniverse.com/dtu/2012/04/monets-ultraviolet-eye.html
22. Radioshack. https://www.radioshack.com/products/radioshack-infrared-led-emitter-and-detector. Accessed from June 2025.
23. Quadros, D. (2013). LED e Fototransistor Infravermelho TIL32 e TIL78, disponível em https://dqsoft.blogspot.com.br/2013/01/led-e-fototransistor-infravermelho.html
24. Lüdke, E. (2010). Um espectrofotômetro de baixo custo para laboratórios de ensino: aplicações no ensino da absorção eletrônica e emissão de fluorescência, *Revista Brasileira de Ensino de Física, 32*, 1506.
25. Cavalcante, M. A., Bonizzia, A., & Gomes, L. C. P. (2008). Aquisição de dados em laboratórios de física: um método simples, fácil e de baixo custo para experimentos em mecânica, *Revista Brasileira de Ensino de Física, 30*, 2501.
26. Lomas, T. C. C., Dickman, A. G., & da Silva Araújo, J. Sequência Didática para o Ensino de doneFísica Ondulatória para Estudantes Cegos, XII Encontro Nacional de Pesquisa em Educação em Ciências – XII ENPEC Universidade Federal do Rio Grande do Norte, Natal, RN, R1431 (8 pages)
27. Toenders, F. G., de Putter-Smits, L. G., Sanders, W. T., & den Brok, P. (2017). Improving physics teaching materials on sound for visually impaired students in high school. *Physics Education, 52*(5), 055006.

# Annex 1

```
#define trigPin 3 // Set trig to 3
#define echoPin 4 // Set scho to 4
#define red 5 // Set red LED to 5
#define green 6 // Set green LED to 6
#define blue 7 // Set blue LED to 7
constintbuzzer = 8;// Set the Buzzer to 8
void setup() {// Pin configuration
Serial.begin (9600); // Starting the serial port
pinMode(trigPin, OUTPUT); // Configure pin 3 as output
pinMode(echoPin, INPUT); // Configure pin 4 as output
pinMode(red, OUTPUT); // Configure pin 5 as output
pinMode(green, OUTPUT); // Configure pin 6 as output
pinMode(blue, OUTPUT); // Configure pin 7 as output
void loop() {
longduration, distance; // Defining parameters for the
ultrasonic sensor
digitalWrite(trigPin, LOW);
delayMicroseconds(2);
digitalWrite(trigPin, HIGH);
// delayMicroseconds(1000);
delayMicroseconds(10);
digitalWrite(trigPin, LOW);
duration = pulseIn(echoPin, HIGH);
distance = (duration/2) * 0.0343;
if (distance< 6){ // Red LED and buzzer lighting
conditions
digitalWrite(red, HIGH);
digitalWrite(green, LOW);
digitalWrite(blue, LOW);
tone(buzzer,1000);// Buzzer activation at 1000 Hz
}
```

```
elseif (distance> 6 &&distance< 30){ // Green LED and
buzzer lighting conditions
digitalWrite(red, LOW);
digitalWrite(green, HIGH);
digitalWrite(blue, LOW);
tone(buzzer,1500);// Buzzer activation at 1500 Hz
}
else { // Blue LED and buzzer lighting conditions
digitalWrite(red, LOW);
digitalWrite(green, LOW);
digitalWrite(blue, HIGH);
tone(buzzer,2000);// Buzzer activation at 2000 Hz
}
delay(500); // Waiting time
}
```

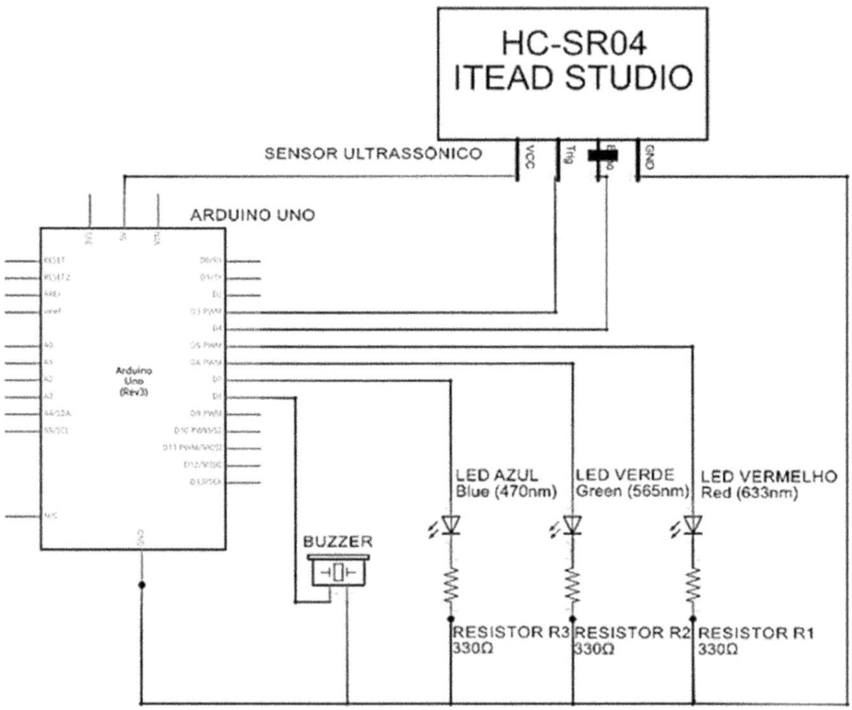


FIGURE A.1 Detailed electronic scheme

# Annex 2

```
#define sensor 2 // Set sensor as 2
#define LED 13 // Set LED to 13
void setup()
{
pinMode(sensor, INPUT); // Configure pin 2 as input
pinMode(LED, OUTPUT); // Configure pin 2 as output
}
void loop()
{
if(digitalRead(sensor)) digitalWrite(LED, HIGH); //
Reads the sensor, and if the status is equal to 1,
turns on the LED and the Buzzer
elsedigitalWrite(LED, LOW); // Otherwise, turn off the
LED and Buzzer.
}
```

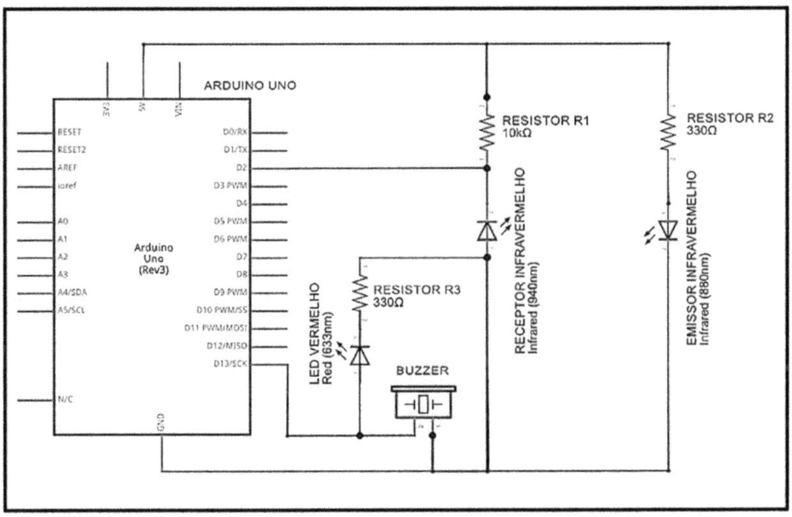


FIGURE A.2 Detailed electronic scheme

CHAPTER 6

# Teaching electrodynamics

## INTRODUCTION

The teaching of science is of fundamental importance for the student's connection with the world and, when it comes to the concepts of electrodynamics, essential for understanding the technological world we live in. Studying science content allows for the development of critical skills, a scientific way of thinking, which can and should be transferred to the students' lives both inside and outside of school. The way in which the student develops this knowledge can be called Scientific Literacy. According to the opinion of Sasseron and Carvalho:

> ...a scientifically literate person does not need to know everything about the sciences (even scientists cannot do that!), but they should have sufficient knowledge of various fields and understand how these studies transform into advancements for society.
>
> *(Free translation from Portuguese) [1]*

The dimensions of scientific literacy should achieve a basic understanding of terms and concepts so that the student can apply them in various situations and in their daily lives, with awareness of the ethical and political dimensions that involve such knowledge and within a perspective of the relationships between science, technology, society, and environment, which denote the benefits, but also the impacts of the use of these technologies [2].

The foundation of Inclusive Education is respect for differences and the guarantee of equitable participation of all students, within the limits of


DOI: 10.1201/9781003510079-6

their abilities. That is, for inclusion to occur, it is necessary to provide a wide range of resources, such as graphs, images, symbols, analogies, and arguments that can be accessed using different channels and sensory skills, as well as the use of appropriate language, as is the case for blind students who use braille for writing and reading [3].

The teaching of science works with modeling phenomena and the use of metaphors and analogies integrated into our senses. However, this does not mean that the absence of a sense capable of perceiving a phenomenon in its essence constitutes the difficulty or impossibility of its understanding. Perhaps the opposite may occur, based on the premise that scientific knowledge does not represent objective reality; often, the "use of visual schemes of phenomena not visually observable can represent conceptual distortions in relation to the knowledge and understanding of these phenomena" [4].

To work with representations of physical phenomena, the teacher usually relies on the blackboard, demonstrations of phenomena, and, less frequently, experimental activities. When it comes to a class with students with disabilities, it is imperative that the teacher has specially developed materials, and preferably tested, for students with visual impairments. These materials can be presented at various levels of complexity and abstraction, with different approaches and through different didactic inscriptions, which are, ultimately, non-verbal representations that can contain one or more meanings, translating into a form of record that carries knowledge produced in research, and can be represented at various levels of detail and abstraction [5]. These inscriptions form a link between the lived world and the conceived world in a continuum. For example, in the lived world, we have measuring instruments such as multimeters, oscilloscopes, and protoboard circuits, which can be represented through photographs or even icons, which retain many characteristics of the photos, up to less detailed representations like diagrams, graphs, and equations [6].

Another concern that should permeate the preparation of the lesson plan when the teacher receives a visually impaired student is the understanding of the students' prior or spontaneous conceptions about the content they intend to work on. A meticulous study conducted by Camargo (2008) concluded that the conceptions of visually impaired students converge toward pre-Newtonian concepts, just like for sighted individuals, meaning that perceptions are not solely obtained through visual stimuli but through the interaction of other senses and social interactions [4].

In electrodynamics, there is an extensive repertoire of the difficulties that students face in understanding and distinguishing concepts such as electric current intensity, potential difference, electrical resistance, electric charge, among others, their interrelations, at various levels of education, from elementary school to higher education [7–10]. With this perception in mind and considering the disparity in the quality of education offered by various institutions, it is necessary for the teacher to seek to understand some details about the journey of visually impaired students for whom they will be teaching.

This kind of anamnesis [11], which usually only occurs after the first evaluations of a class, should be preceded if the teacher is aware that they will receive a student with visual impairment. Initially, the opportunity to talk with the guardians and former teachers can help understand the adaptations the student is familiar with, the level of depth worked with this student, the barriers that have not yet been overcome, among other questions inherent to the necessary prior knowledge. One should also interview the student to better understand the assistive technologies they are proficient in, such as reading and writing in Braille, their mastery of prior knowledge for the content to be covered, among other particular needs, such as the need to sit closer to the teacher to follow the class, since, without access to the board, hearing the teacher's voice loud and clear is imperative.

Only based on the collected data can the teacher develop the Personalized Learning Models (PLM), acquire or create adapted materials, such as tactile representations that can be used by students with visual impairments, but preferably also designed to capture the attention of other students [12].

This entire process may seem very laborious, and it truly is [13]. At least when the teacher starts receiving students with disabilities, with practice their efficiency tends to increase. Thinking about inclusion without taking all these precautions means not promoting true Inclusive Education. Despite a teacher who teaches 20, 40 different classes per week with 30–50 students each and without a paid workload specifically to assist visually impaired students, or even students with various disabilities, in an institution without structure and financial or auxiliary support, one cannot speak of inclusion, at most integration [13, 14].

Although the objective of this book is to provide content that can serve as a manual for teachers who need to teach physics to a class with visually impaired students, we find it appropriate to emphasize the need to value this professional who, ultimately, will generate and manage Inclusive Education in their classroom [14].

Over time, while performing their activities, the teacher acquires a body of knowledge in their teaching practice that Camargo called Teacher Knowledge [13], necessary for working with students with visual impairments. Among the knowledge, we can mention: critical mastery of the content; breaking away from simplistic views on science teaching; questioning "common sense" ideas about science teaching, becoming critical of traditional teaching; developing activities that produce effective learning; knowing how to guide students to learning; knowing how to evaluate; staying connected to research in teaching in order to improve their didactic practice; and the concern with communication and language, which in the case of blind students is associated with tactile and auditory representations [13].

Camargo [13] analyzed the difficulties and feasibility of teaching electrodynamics to visually impaired students and pointed out that, regarding the difficulties, the main problems are related to the meanings linked to visual representations, whose inherent difficulties in communication, mathematical operation, and communication end up being the cause of segregation. As for the possibilities (ways to deal with the difficulties that produced positive results), within a predominantly interactive methodological context, it includes: the presentation of models and hypotheses fundamentally auditory; auditory and auditory-tactile communication; and the use of experiments in tactile and auditory forms [13].

Therefore, we realize that the teacher's concern regarding teaching a class with visually impaired students should be auditory communication, since these students will not have access to gestures and the blackboard, as well as the use of tactile representations that can, not replace, but complement this teacher's repertoire in order to reach students with disabilities, also bringing benefits to the understanding of the phenomenon studied by the other students.

Effectively, the discipline of physics still does not have a large range of research articles involving teaching for visually impaired people [14], compared to other disciplines in the STEM fields. Still, in a review conducted by Castro-Castillo and Tuay-Sigua [15], Brazil stands out in its research, contributing about 70% of the published articles from 2010 to 2022, aimed at teaching for people with visual impairments.

## METHODOLOGY

The challenge of developing conceptual thinking was deeply studied by Vygotsky, which in simplified terms would be thinking by concept, knowing how to relate a word to its meaning, that is, the study of the path that leads a child to develop these concepts in their mind, which will be important when we think about concepts related to physics. Vygotsky categorized three phases for the development of the child: the first phase corresponds to clusters or syncretic aggregates, the second, longer phase, complex thinking, until reaching the final stage that completes mental structures, called the phase of potential concepts [16].

Even though for Vygotsky, just like for Jean Piaget, the genetic structure of the human brain is completed in adolescence, this does not mean that this structure is in use, since many adolescents and adults have not yet reached or continue using complex thinking, which would be to be satisfied with answers based on simple correlations that are part of the normal development process, but that distance themselves from the acquisition of conceptual understanding [16]. One of the examples used by the author, which leads to pseudoscientific concepts, is the fact that stars, which have their own light, are seen in the sky twinkling, unlike planets which maintain a fixed brightness. This could lead to the conclusion that, because they emit their own light, stars twinkle, showing an implausible cause-and-effect relationship since stars appear to twinkle to us simply because they are much farther from Earth than planets [16].

Vygotsky's studies involved innovative thoughts on disability, favoring a qualitative view over a purely quantitative perspective, rejecting the idea of disability being represented solely by loss or limitation, advocating for its compensation through other psychological and social functions, that is, disability should not be seen merely as a deficit or incapacity, but as a restructuring factor in the child's development, creating another path for their development through compensatory processes and reconfiguration of psychological functions [17].

According to Vygotsky, human development is governed by a biological (natural) line and a cultural-historical line that involves social interaction and the appropriation of cultural tools, such as language, which becomes even more fundamental for the child with disabilities, given the crucial need for social interaction and mediation with the environment for the development of their potential [17].

Vygotsky criticized purely medical and biological approaches, seeing the school as a fundamental factor of inclusion that allows integration with peers and access to pedagogical and psychological tools capable of transforming "difference" into an advantage depending on the context and strategies used, enabling children to develop within their capacities. In summary, in Vygotsky's optimistic view, centered on the idea of social interaction, education and the cultural environment are fundamental for the child's development [18].

Therefore, we believe that a methodology based on Vygotsky's thoughts is preferable for promoting Inclusive Education within an approach that allows interaction between visually impaired students and other students, mediated by the teacher who must always take their potential into account, helping to promote the elimination of social barriers that limit their participation in the teaching and learning process.

## ANALOG TACTILE-VISUAL REPRESENTATIONS

We will present articles with proposals for the development of materials that can assist in the teaching of electrodynamics for visually impaired students. Most of the developed materials consist of models or tactile-visual representations of models that represent physical concepts, but tactile-auditory models and instruments are also used [13].

In [19], tactile-visual models representing negative and positive charges (respectively, electrons and holes) were developed. One charge was represented by a Styrofoam ball and the other by a square-shaped etilene vinil acetate (EVA) rubber sheet, with a central circular hole the same diameter as the Styrofoam sphere, with the aim of showing that only different charges will fit together, which is related to the law of attraction of opposite charges.

The article [20] also used EVA, but this time representing, with this same material, the negative and positive charges. The negative charges, the electrons, are this time represented by the circular cut made in the center of EVA sheets cut into square shapes. When the materials do not fit together, it means they are of the same type of charge, that is, both positive or both negative. And when the materials fit together, it indicates that they are opposite charges attracted to each other, tending to stay close.

To represent an electrical circuit, the movement of charges associated with electric current, the different electric potentials, and the effect of resistance on the movement of these charges, a study [19] used an analog tactile-visual model also made of EVA foam, where paths made of grooves in this material allow the movement of metal balls. The entire path is on the same plane, except for the source, which is represented by two small semicircular elevations of different sizes that are associated with two different potentials, positive and negative. At certain points along the paths formed by the grooves, there are pieces of felt that hinder the movement of the balls in that section, representing electrical resistance. On the elevations, felt was also applied to represent the internal resistance of the source. The same material also allowed for the representation of series and parallel resistor associations.

In another work [21], we found a proposal for an analog tactile-visual model that closely resembles that of Figure 6.3. This model is also made of EVA and features two semicircles that define the positive and negative potentials, made of Styrofoam and in different sizes. The main characteristic that differentiates it from Figure 6.3 is that, in this model, a strip of double-sided tape was placed along the circuit, and on the tape, several marble balls (glass balls) were adhered to the electrons.

The article [22] presents diagrams of electrical circuits containing an association of three resistors in parallel and in series, connected to a power source, using plastic straws and EVA. The use of Braille writing indicating each element of this tactile-analog representation is noticeable. For the study of conductivity, where the student should understand the concepts of conductor and insulator, we found a remarkably interesting experiment [23] that allows the student with visual impairment to discover if the material is a conductor or insulator by pressing a button. When the student presses the button, the circuit can be closed if the material that makes up the circuit is a conductor of electricity. If the material is conductive, pressing the button closes the circuit and activates a sound device (buzzer).

One of the tasks assigned to high school and college physics students is the creation of graphs. This construction, for the visually impaired student, can be carried out in various ways, as described in [21]. In this work [21], a transparent PVC film (a type of acetate) was used, molded in a device called Thermoform, which employs heat and vacuum to shape this film from a mold. In this case, the film was molded as a Cartesian plane, formed by a raised grid pattern with about 196 small squares, analogous to graph paper. In the center of the figure, there are two perpendicular axes with arrows at the ends that intersect in the center of the figure, forming the four quadrants of the Cartesian plane. Using this material, the student can build and rebuild graphs using pieces of string and adhesive tape. The advantage of this model is that it allows for the inscription of both negative and positive values on the x-axis as well as the y-axis. However, the task of gluing the string does not seem very easy to us, if we imagine that the student will have to do it alone.

Another possibility is described in [24] using magnets on a metal base for the construction of graphs. The idea is very interesting because it involves low-cost materials, as a thermoforming machine can cost several thousand U. S. dollars and it allows the student to carry out the entire procedure without the help of another person. Small neodymium magnets are used, as well as strips of magnets commonly used in refrigerator door seals and self-adhesive magnetic sheets used in the production of refrigerator magnets.

In a study [25], the potential difference was associated with an inclined plane down which small balls representing electric charges rolled, and then obstacles (nails) were added to the ramp to represent electrical resistance. So that the student could develop the experimental practice, they created a protoboard using banana jacks (or sockets) and electronic components, such as resistors and wires attached to plates made with 3D printers and with the terminals soldered to banana plugs. To measure current and potential difference in the circuit, a device was developed using the Arduino prototyping platform that allows the student to obtain the results audibly.

## TEACHING ELECTRODYNAMICS

The electrodynamics course generally begins with the initial concepts about atomic models, where various types of tactile representations can be made available, such as those developed in [26], using Styrofoam balls to represent the atomic model of the hydrogen atom, positively and negatively charged atoms, and crystalline structures also made with Styrofoam and popsicle sticks, representing the weak bond that occurs in metals that are good conductors of electricity. The manipulation of these tactile-visual representations would be useful and very welcome for the entire class, within an inclusive lesson, as it allows for a playful and three-dimensional visualization of something that students are used to seeing only drawn in their books or on the blackboard. It is expected that, with these activities, the student will be able to master the understanding of the concepts of atomic models (composed of protons, neutrons, and electrons), electric charge, positive charges (absence of electrons) and negative charges (excess of electrons), metallic crystalline structures, and free electrons. Figure 6.1 shows a blind student handling the materials.

An important tactile-visual representation developed in [26] can be used to represent the movement of electric charges through a circuit composed of a switch, a potential difference, and two resistors connected in parallel, as shown in the following figure (Figure 6.2). Metal balls, representing electric charges, are pushed by the student's fingers through a channel made of EVA with a base made of fiberboard the student lifts the

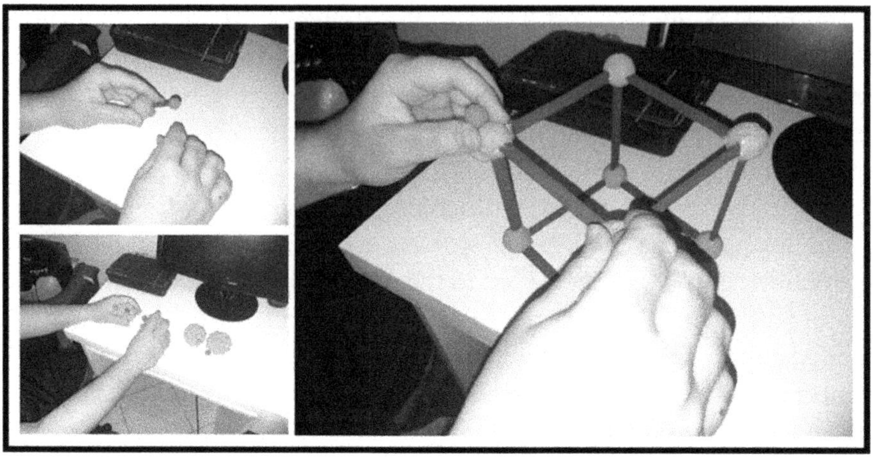

FIGURE 6.1 In the first image at the top left, the blind student manipulates a tactile-visual representation of the atomic model of the hydrogen atom. In the figure below and to the left, we see the manipulation of tactile-visual representations of atoms with an excess of electrons and atoms with a deficiency of electrons, which fit together simulating an atomic bond. In the figure on the right, the student manipulates a tactile-visual representation of a crystalline structure of atoms, where there is a free electron that weakly binds to the atoms through magnets placed inside both. The materials used were Styrofoam balls and popsicle sticks. The colorful painting serves, in an inclusive class, to spark greater interest for the students who can see. The colorful painting serves, in an inclusive classroom, to spark greater interest for the students who can see.

switch, composed of an obstacle also made of EVA, which prevents the metal spheres, representing a circuit with the switch off. When the student lifts the obstacle, opens the switch, they can push the balls (which represent the charges) that can move to the right through a wide path, without difficulties, which represents the movement of charges through a conductor wire. To the right, following an anti-clockwise path, there is an elevation that takes the spheres to the highest part of the circuit. This elevation represents a potential difference, and this path is a bit narrower, with the spheres passing through with very little difficulty, which represents the internal resistance of the battery. Upon reaching the highest point, at the curve, the ball passes without difficulty, which represents a conductive wire. Right after that comes a straight, narrow path with a slope, where the balls pass with some difficulty, as this path represents an electrical resistor. The slope indicates the potential drop across the

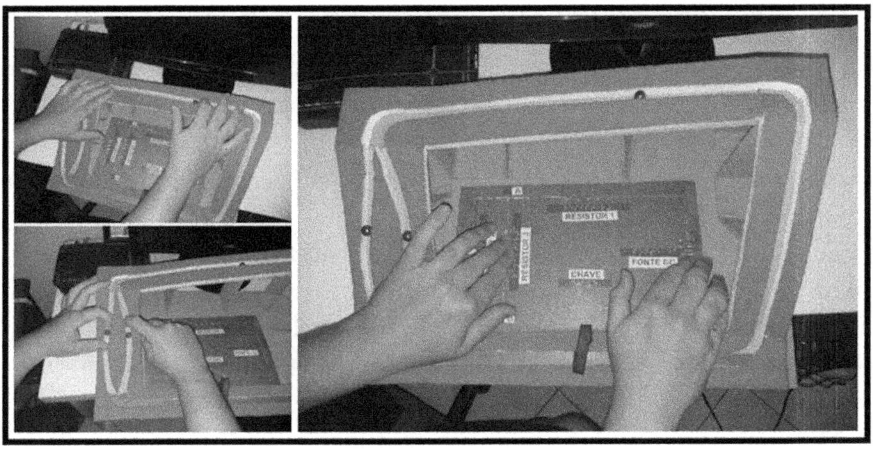


FIGURE 6.2 In the first image at the top left, the blind student manipulates a tactile-visual representation, pushing the metal ball through the path that represents the first resistor of the circuit. In the lower left image, the student manipulates two spheres with both hands, simultaneously along the two paths that represent the two resistors in parallel. In the image on the right, the student manipulates the tactile-visual representation of the circuit, on the inner part, in relief, using touch to understand the symbols corresponding to the components and how the movement of charges works and the functionality of each component.

resistor. Once again, a wide curve follows, through which the balls pass freely, representing a conductor wire that connects the first resistor to two more. These last two resistors are represented by paths placed side by side, where the spheres pass with difficulty in a final slope that leads the balls back to the starting point, indicating two resistances connected in parallel [26, 27].

On the inner part of the tactile-visual representation of the electrical circuit, there is a contiguous representation, where all the circuit components were drawn with rubber glue, identified with their symbols, and Braille labels were inserted indicating the name of each circuit component. To make the tactile-visual representation attractive to sighted students, labels with the names of the components were applied in ink, and the entire apparatus was painted in attractive colors [26].

Using the proposed material, it is therefore possible to develop various introductory concepts such as the concept of electric charge, electric current, electrical resistance, conductor, potential difference, resistor association, among others.

## EXPERIMENTAL APPROACH

Laboratory activities are generally much more attractive to students since they can take on a more prominent role than in traditional classes. Hands-on activities help to complement the chalkboard lesson, but they can even be used first depending on the objective, allowing students to participate in a science-making activity.

For these activities, it has always been very difficult to consider the possibility of a blind student participating. However, articles [26, 27] provide us with a wide variety of materials that can contribute not only to the blind student but to the entire class, with a simple methodology, but with various possible levels of depth and a lot of playfulness.

In articles [26, 27], two adapted multimeters were used with Braille inscriptions next to the selector and on the probes, which, in addition to labels indicating polarity, also have magnets to attach to ferromagnetic surfaces. In Figure 6.3, we can observe above and to the left, a box containing various everyday materials, such as cardboard, plastic bottles, glass bottles, pencils, screwdrivers, among many others. In addition to these materials, there are batteries and boards containing electrical components (resistors and conductive wires). The initial goal of the student is to test, using the multimeter on the conductivity test scale, which materials conduct electricity and which do not. Placing the multimeter on the

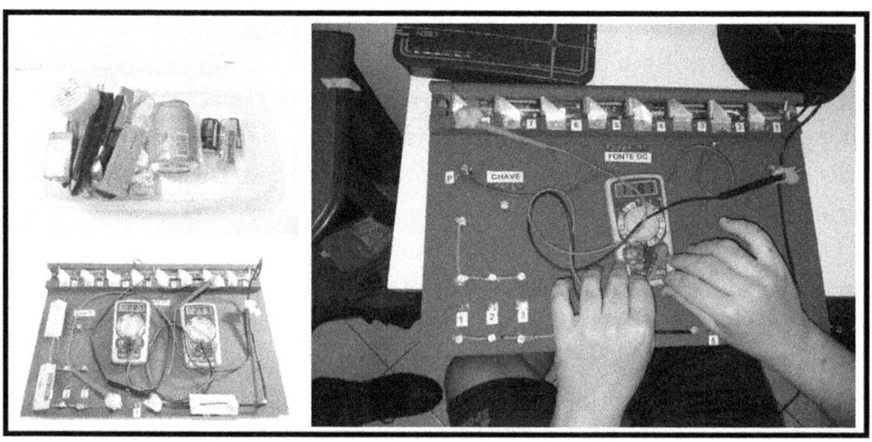

FIGURE 6.3 In the first image at the top left, we have a box with various everyday objects, including batteries and cells, as well as electronic components attached to a wooden board and adapted to magnetic screws. In the lower left image, we have a protoboard adapted to receive magnetic electronic components and test probes. In the image on the right, the blind student manipulates the protoboard.

conductivity test scale, when the material is conductive, a beep sound is emitted. Initially, the teacher can help the student by indicating what the object is and asking the student, before taking the reading, whether the material is conductive or non-conductive. Then you can let the student try to identify the material the object is made of and attempt to predict, before testing with the multimeter, whether the item is conductive or not. These activities are especially important since, without the sense of sight, the student becomes much more vulnerable to the possibility of an electric shock in everyday situations.

The boards with electronic components have magnetic terminals that attach to the multimeter probes, allowing the student to connect them and keep their hands free to handle the multimeter's resistance measurement scales. Thus, the student can use the equipment to take measurements independently, but still, in this case, depending on someone who can see to read the result on the display. A multimeter with Bluetooth can be used, which, when connected to a cell phone, allows the blind student to even read the measurements. However, the cost of this equipment is very high compared to the multimeter used in articles [26, 27]. On the continuous potential difference scale, the student can read the provided batteries and cells, also making connections through the magnetic probes.

In the lower left side of Figure 6.3, we see a protoboard specially developed to facilitate circuit assembly and readings, with the adapted multimeter. In this protoboard, there is a set of C-type batteries, connected in series with braille indicators and magnetic connections that allow the connection in multiples of 1.5 volts, depending on how many batteries are connected. Magnetic screws, some connected to each other, allow the assembly of circuits in different configurations (series, parallel, or mixed resistor associations), which, with the use of the adapted multimeter, also allows the student to simultaneously measure current and voltage drop across points in the circuit. In this way, for example, the student can take readings of voltage and current for different potential differences, which allows them, with the collected data, to construct a graph, as we will see next. In the image on the right, the blind student handles the protoboard Figure 6.3.

It is important to highlight that all the equipment described above can be used by other students, allowing for greater practicality in circuit assembly and reading measurements, in a playful and fun way, enabling the student to "follow" the path of the currents with their fingers or eyes, unlike traditional protoboards, where many of these connections remain hidden.

With the readings obtained in the previous activity, where the current and voltage drop were measured across the same resistor, the student can create the graph with the recorded points by placing the current values on the x-axis and the voltage drop values on the y-axis. In articles [26, 27], we see the construction of a device that allows blind students to create graphs tactilely (Figure 6.4). The device consists of a metal frame, designed to hold photos with magnets, onto which an A3-sized graph paper sheet was attached. On top of this sheet, a transparent acetate sheet was placed. Using a craft knife and ruler, the acetate was scored over all the lines of the A3 paper, forming a tactile version of the paper on the surface of the acetate. On the axes, sewing lines were glued so that the student could perceive the tactile difference of this line (greater relief) from the others (scratched with a scalpel). On the axes, labels in Braille with the numbers of the two axes were affixed every centimeter. After defining the scale, the student can mark the obtained points with 5 mm neodymium magnets and adjust the best line with a bar magnet used in refrigerator doors. In this way, having the slope of the best fit line, the student can calculate the resistance of the resistor and compare it with the reading taken with the multimeter, which are practical lessons for understanding Ohm's first law.

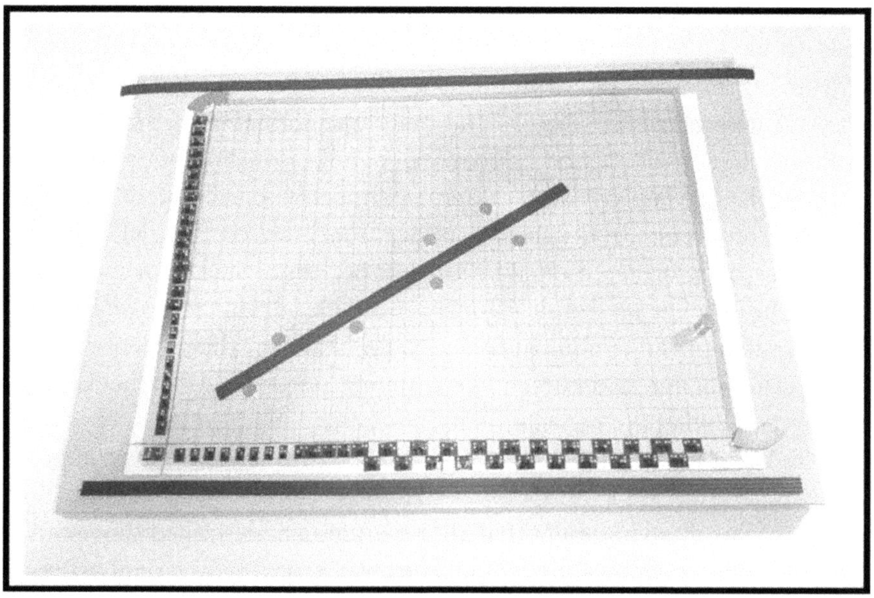

FIGURE 6.4 The apparatus developed for the construction of tactile graphs, using magnets.

A major difficulty for students is understanding images, even when they are transformed using specific programs into braille dots or turned into models. In assessments, for a class where there are students with visual impairments, the images present in the assessment need to be accessible to these students, and there is no consensus on the best way to do this. Generally, instead of the figure, its description is provided in words, which can become a significant obstacle. The artificial intelligence (AI) technologies currently developed can read images, but as shown in article [28], their performance is still not satisfactory. The article [28] researched the use of ChatGPT-4 to assist in the description of figures and physical concepts involved and found that the description is imprecise, advising teachers to exercise caution in using it for educational purposes, including as an accessibility tool for blind and visually impaired students. However, with the evolution of these AI technologies, we foresee a different scenario over the years.

## COMMENTS

Electrodynamics is a field of physics that has many studies attempting to understand the reasons behind students' conceptual difficulties. Many other studies provide contributions in the form of didactic materials and the development of different teaching modality proposals. The challenge is amplified when we receive a visually impaired student, given the degree of abstraction of the concepts and the visual representations we are accustomed to using to teach this area of physics. The articles discussed in this chapter contain tactile-visual representations and other materials that can be made at a relatively low cost, allowing the teacher to adapt them, based on the initial ideas and principles of operation, according to the student body they are teaching. We encourage teachers who do so and who produce diverse materials with positive learning experiences to make their ideas public. Given the very limited number of materials for teaching physics to visually impaired students, the teaching of physics lacks good ideas to better serve everyone.

## BIBLIOGRAPHY

1. Sasseron, L. H., & Carvalho, A. M. P. de. (2011). ALFABETIZAÇÃO CIENTÍFICA: UMA REVISÃO BIBLIOGRÁFICA. *Investigações Em Ensino de Ciências, 16*(1), 59–77. https://ienci.if.ufrgs.br/index.php/ienci/article/view/246. Accessed on April/22/2025.
2. Silva, M. B., & Sasseron, L. H. (2021). ALFABETIZAÇÃO CIENTÍFICA E DOMÍNIOS DO CONHECIMENTO CIENTÍFICO: PROPOSIÇÕES PARA UMA PERSPECTIVA FORMATIVA COMPROMETIDA COM

A TRANSFORMAÇÃO SOCIAL. *Ensaio Pesquisa Em Educação Em Ciências (Belo Horizonte)*, 23. https://doi.org/10.1590/1983-21172021230129. Accessed on April/22/2025.

3. Diana-Carolina Castro-Castillo, & Rosa-Nidia Tuay-Sigua. (2023). Revisión sobre la enseñanza de la física en estudiantes con diversidad funcional visual. *Revista Científica*, 46(1), 1–16. https://doi.org/10.14483/23448350.17577. Accessed on April/22/2025.
4. Camargo, E. D. (2008). *Ensino de Física e deficiência visual: dez anos de investigações no Brasil*. São Paulo: Plêiade/FAPESP.
5. Alves, E. G. (2011). Um estudo multimodal de textos didáticos sobre o efeito fotoelétrico, Dissertação de Mestrado, Universidade Federal de Minas Gerais, Belo Horizonte, MG.
6. Roth, W.-M., Pozzer-Ardenghi, L. E., & Han, J. Y. (2005). *Critical Graphicacy: Understanding Visual Representation*. Dordrecht: Springer.
7. Mcdermott, L. C. (2014). Melba Newell Phillips medal lecture 2013: discipline-based education research: a view from physics Lillian Christie McDermott citation. *American Journal of Physics, 82*, 729.
8. Lederman, N. G., & Abell, S. K. (2014). Teaching Physics. *Handbook of Research on Science Education* (Vol. II). Ed. Norman Lederman. New York: Routledge, 439–445.
9. Duit, R., & Von Rhöneck, C. (1997, 1998). Learning and Understanding Key Concepts of Electricity From. *Connecting Research in Physics Education with Teacher Education*. International Commission on Physics Education.
10. Arons, A. (1996). *Teaching Introductory Physics*. Compilation of 3 previously published works. John Wiley and Sons, 188–189.
11. Mendonça, A. J. B., Maciel, & A. K. S. (2018). Anamnese – Etapa que Antecede a Elaboração de Um Programa de Orientação e Mobilidade. *Orientação e Mobilidade: ensaios. Ministério dos Direitos Humanos*. Ed. Freitas, L. P. T., & Façanha, A. R. Projeto de Especialização e Mobilidade. Instituto Federal de Educação, Ciência e Tecnologia do Ceará (IFCE), 35–45.
12. Rhim, L., & Lancet, S. (2018). How Personalized Learning Models Can Meet the Needs of Students with Disabilities: Thrive Public Schools Case Study. https://files.eric.ed.gov/fulltext/ED582803.pdf. Accessed on April/22/2025.
13. Camargo, E. P. (2012). *Saberes docentes para a inclusão do aluno com deficiência visual em aulas de Física*. São Paulo: Editora Unesp.
14. Silveira, M. V. da, & Santos, A. C. F. dos. (2022). percepção dos professores de Física sobre a inclusão de estudantes com deficiência visual: uma pesquisa quantitativa. Revista Educação Especial. https://doi.org/10.5902/1984686x66730. Accessed on April/22/2025.
15. Castro-Castillo, D. C., y Tuay-Sigua, R. N. (2023). Revisión sobre la enseñanza de la física en estudiantes con diversidad funcional visual. *Revista Científica, 46*(1), 1–16. https://doi.org/10.14483/23448350.17577.
16. Gaspar, A. (2014). *Atividades experimentais no ensino de física: uma nova visão baseada na teoria de Vigotski*. São Paulo: Editora Livraria da Física, 209.

17. Vygotsky, L. S. (1987). *The Collected Works of L. S. Vygotsky: The Fundamentals of Defectology (Abnormal Psychology and Learning Disabilities)*, Translated and with an Introduction by Jane E. Knox (Vol. 2). New York: Kluwer Academic/Plenum Publishes, 97–112.
18. Vigotski, L. S. (2022). *Obras Completas - Tomo Cinco: Fundamentos de Defectologia./Tradução do Programa de Ações Relativas às Pessoas com Necessidades Especiais (PEE).* Cascavel, PR: EDUNIOESTE, 488 p.
19. Souza, M. M., Costa, M. P. R., & Studart, N. (2008). Tecnologia para o ensino de eletrodinâmica para o aluno cego. *Física na Escola, 9*(2), 10–13.
20. Silva, M. R., & Pierson, A. H. C. (2013). Ensino de física e deficiência visual: relato de uma experiência em aulas de eletrostática, XX Simpósio Nacional de Ensino de Física - SNEF 2013 - São Paulo, SP.
21. Andrade, L. M. (2017). Inscrições didáticas para um estudante cego em uma unidade didática de circuitos elétricos resistivos, Dissertação de Mestrado, Universidade Federal de Lavras, MG.
22. Filho, M. B. M. (2015). O ensino de eletromagnetismo para alunos com deficiência Visual, Dissertação de Mestrado, Universidade Federal de Goiás, Catalão, GO.
23. Almeida, L. C., Xavier, C. T. A., & Marinho, K. S. O. (2012). Ensino de física e educação inclusiva: exemplo de uma sequência didática para a abordagem de conceitos de eletrodinâmica. *Ensino, Saúde e Ambiente, 5*(2), 102–113.
24. Azevedo, C. A., & Santos, A.C. F. (2015). Ensino de Gráficos para que alunos cegos possam fazer interpretações de fenômenos físicos. *Caderno Espaço Inclusão, 3*(1), 56–64.
25. Andrade, H. B., Schwerz, R. C., & Sabo, P. H. (2023). Eletrodinâmica: atividades experimentais multissensoriais visando a inclusão de alunos com deficiência visual. *ACTIO: Docência Em Ciências, 8*(2), 1–19. https://revistas.utfpr.edu.br/actio/article/view/16427/9648
26. Velloso, M., Arana, M., Acioly, V., & Santos, A. C. F. (2021). Abordagem da Lei de Ohm para Estudantes com Deficiência Visual no Ensino Superior. *Latin-American Journal of Physics Education, 15*(1), 10. https://dialnet.unirioja.es/descarga/articulo/7918018.pdf. Accessed on April/22/2025.
27. Velloso, M., Arana, M., Acioly, V., & Santos, A. C. F. (2021). Hands-on electricity remote teaching to a blind student during pandemic of 2020. *Physics Education, 56*(5), 055027. https://doi.org/10.1088/1361-6552/ac10bb. Accessed on April/22/2025.
28. Polverini, G., Melin, J., Onerud, E., & Gregorcic, B. (2024). Performance of ChatGPT on tasks involving physics visual representations: the case of the brief electricity and magnetism assessment. ArXiv (Cornell University). https://doi.org/10.48550/arxiv.2412.10019.

CHAPTER 7

# Teaching optics

## INTRODUCTION

Visual impairment affects the eyes, not the intellect. Visually impaired individuals possess the same capacity for learning and curiosity as those without visual impairment. Despite their inability to see the world, they share a desire to comprehend it. Therefore, the teaching of physics should accommodate individual needs. In addition to challenges with written material, visually impaired students often struggle with diagrams and charts that are difficult to grasp through oral explanations alone. Since tactile sensation is key for the blind to perceive their surroundings, experiments should be designed to engage their sense of touch.

While studies have explored the challenges encountered by individuals with physical disabilities in learning science [1–11], research on teaching physics to visually impaired students remains limited. One common oversight is the reliance on teaching models designed primarily for sighted individuals, neglecting the unique needs of blind students.

In the conventional classroom setting, a significant challenge in teaching physics to visually impaired students stems from their limited vision, often resulting in a lack of comprehension. Without clear visual cues, there's a heightened risk of misconceptions arising from the information presented. Recognizing this, the authors advocate for a structured approach to teach optics to visually impaired students, emphasizing the central role of models. It's crucial that instructors are well equipped to implement inclusive education practices tailored to the needs of visually impaired students.


DOI: 10.1201/9781003510079-7

Despite the longstanding development of electronic devices to aid visually impaired students, a significant portion of the global student population lacks access to such technology [8]. Alternative materials have been proposed in the literature, including frames with raised edges filled with wax, allowing students to write or draw with the assistance of a pointed tool [8]. In this chapter, our focus is primarily on the challenges visually impaired students encounter when reading graphics and diagrams. We propose practices utilizing readily available, low-cost materials that offer tactile feedback to address these difficulties effectively.

Specialized schools traditionally catered to students with special needs. However, the prevailing perspective now advocates integrating students with special needs into regular education settings alongside their peers. Yet, this shift necessitates a prepared educational system that accommodates diverse needs. For visually impaired students, this entails strategic teaching approaches, adapted instructional materials, and modifications to the standard curriculum to prioritize tactile, auditory, and communicative skills. Visually impaired students face challenges when conducting experiments, particularly in understanding optics where forming a mental image of light is crucial. This chapter also offers practical alternatives for experimental work and contributes to developing a conceptual understanding of light, recognizing that our knowledge of nature stems from sensory input. While sighted individuals rely on vision, for the blind, other senses like hearing and touch take precedence. Traditionally, teaching optics to visually impaired students has involved tactile visual aids, but these have limitations due to the absence of vision. To address this, it's essential to design experiments that simulate real experiences as closely as possible. Many optical experiments rely heavily on visual observation, so adaptations are needed to translate light signals into formats accessible to blind students. Touch, a primary sense for the blind, plays a crucial role in this adaptation, as explored in this work. We advocate for teachers to utilize model-building approaches to engage students actively, fostering reasoning skills and deeper comprehension of concepts.

In addition, this chapter proposes an unconventional application of lasers, emphasizing the importance of adhering to safety standards to mitigate the risk of accidents, particularly those involving eye injuries. Even at relatively low intensities, laser light can inflict permanent damage to the eyes. Moderate to high-powered lasers pose an even greater threat, capable of causing burns to the retina or skin. Regulations such as ANSI Z136 in the United States [12] and the international standard IEC 60825 [13] classify lasers into different 'classes' based on their power and wavelength, outlining safety protocols including specific labeling and the mandatory use of safety goggles during laser operation. These precautions are crucial, especially for visually impaired individuals who may lack the reflex response to rapidly close their eyes when exposed to intense light.

## STUDENTS' MISCONCEPTIONS IN GEOMETRICAL OPTICS

Students approach optics with a naïve conception of light and its properties. All students know that light travels in a straight line, but few have a clear understanding of what this means. Many students don't recognize that light is a physical entity, with an existence apart from its sources and effects. They don't make a clear distinction between light and vision.

## DIFFICULTIES IN DESIGNING OPTICS TEACHING ACTIVITIES FOR STUDENTS WITH VISUAL IMPAIRMENTS

Camargo and Nardi [14] present results from a study analyzing the performance of prospective physics teachers. During a teaching practice course, these future educators were tasked with planning, developing, and delivering optics lessons to a class that included visually impaired students.

A significant challenge identified by Camargo and Nardi [14] was the reliance on visual observation for understanding optical phenomena. The prospective teachers initially struggled to decouple the concept of "knowing" from "seeing," which is crucial when teaching students with visual impairments.

Another challenge is the adherence to traditional pedagogical methods. Camargo and Nardi [14] observed a tendency among the future teachers to adhere to conventional teaching methodologies. This traditional approach often posed barriers to creating inclusive learning environments suitable for visually impaired students.

Camargo and Nardi [14] suggest some strategies and alternatives to be developed by the teacher: Development of Tactile Materials: To address the challenges, the prospective teachers demonstrated creativity by developing tactile teaching aids. These materials enabled students with visual impairments to engage with and understand optical concepts through touch.

Another suggestion is to put emphasis on oral communication. According to Camargo and Nardi [14], the educators incorporated detailed verbal descriptions and discussions into their lessons, reducing the reliance on visual demonstrations. This approach facilitated a more inclusive classroom environment, accommodating the needs of all students.

The authors also investigate communicative strategies for the inclusion of students with visual impairments in physics education, with a focus on optics topics [15]. Camargo and Nardi analyze the feasibility of interactions between future teachers and students with visual impairments, highlighting the importance of the empirical and semantic-sensory structures of the languages used.

The authors make some recommendations. The teacher should identify the semantic-sensory structure of conveyed meanings, i.e., to understand how students with visual impairments perceive and interpret optical concepts is essential for effectively adapting teaching language.

The teacher should also know of the student's visual history, i.e., to understand each student's visual background allows educators to personalize teaching approaches that meet individual needs, considering previous experiences and the degree of remaining visual perception.

The teacher should make use of tactile-auditory empirical language in interactive contexts. Integrating tactile and auditory resources in an interdependent manner facilitates the comprehension of optical phenomena, promoting a richer and more accessible learning experience.

Finally, the teacher should exploit the communicative potential of fundamental auditory and auditory-visual empirical languages. Maximizing the use of auditory resources and, when possible, visual ones independently expands communication possibilities and enhances students' understanding of the content.

Camargo and Nardi [14, 15] conclude that, although there are significant challenges in adapting optics teaching to visually impaired students, it is possible to overcome them through creativity and the willingness of teachers to develop and implement inclusive pedagogical strategies. The experience highlights the importance of preparing future educators to deal with diversity in the classroom, promoting an educational environment that is accessible to all students. Camargo and Nardi [14, 15] emphasize the need to adapt pedagogical practices to ensure accessibility and the effective participation of students with visual impairments in optics education. By considering the proposed recommendations, educators can foster a more inclusive and effective learning environment.

## MEDIATED LEARNING

One of the concerns expressed by regular education teachers when dealing with visually impaired people in their classes is associated with the nature of the student's instruction and the resources available for this learning. Modern educational trends suggest a form of teaching where the student has an effective participation in learning. They all recognize the great importance of the student's initial knowledge, that is, the knowledge that the student brings with them, and the need for self-evaluation on the part of students. However, the different theoretical frameworks differ very significantly regarding their expectations regarding the student's prior knowledge. Some believe that students should be left completely free during the learning process, while others believe that this strategy is inefficient.

An alternative is mediated learning, a form of student-centered instruction that does not allow students to wander through the learning process alone but introduces some structure into the process. The idea is to engage students in the construction and development of a model and the formation of concepts in accordance with the standards of scientific research. A concept is understood as the "representation of an object by thought, through its general characteristics" [3], and a model is an abstract and simplified view of a complex reality. The term conceptual model can be used to refer to models that are represented by concepts or related concepts that are formed after a process of conception in the mind. A mental model can be understood as a representation of something in the mind.

Interactive engagement methods are among the most meaningful teaching methods. One way to get students to actively participate is to adopt mediated learning, that is, through intense interaction between the student and the teacher, or mediator.

As a matter of adaptation, our impressions about the world around us manifest themselves most visually. Vision is responsible for at least 80% of the information acquired by humans [4]. Thus, in the case of visually impaired students, several questions arise: How will the student understand the concepts presented in class? How will you understand the concept of light? How to describe the phenomena of refraction and reflection of light?

## BUILDING MODELS ABOUT LIGHT

It is essential to highlight that one of the ways blind students interact with the world is through touch. Touch is the main source of information about the mental representation of an object for those without visual stimulation. There are differences between the perception and processing of information acquired through touch and that acquired through vision. The acquisition of information through touch takes longer compared to that acquired through vision, in addition to having a sequential nature, generating a greater load on working memory [7]. However, active touch must be differentiated from passive touch. In passive touch, information is received unintentionally. On the other hand, in active or haptic touch, information is intentionally sought by the subject [7]. The visually impaired person must face many almost insurmountable obstacles in learning science, especially physics. One of the main reasons is the lack of material adapted for use in physics classes.

In a conventional class, students are forced to readily assimilate everything the teacher says. They must be able to decipher and organize, coherently and on their own, any information presented to them, regardless of the form in which it is presented.

Teachers often provide students with simplified mental images for physical systems under study, for example, a constant acceleration of gravity in the vicinity of the Earth's surface. These images are called working models for the system. A working model is an abstraction of reality [5]. The human mind is not capable of understanding all the complexity of all the details of a real system. A working model constitutes an idealized portrait. No model fully corresponds to reality, and we do not know a priori the limits of a model. In fact, the concepts of right and wrong do not apply to models. We can at most classify a model as suitable or not suitable. The teacher must check whether a given model is suitable for his or her purpose.

A typical physics class usually requires the use of diagrams and graphs. But how should we make the drawing so that the visually impaired student can feel it tactilely? Some criteria must be followed: the material must be easily obtained [8–16], lines, diagrams, and drawings must be easily erased or removed, and allow the student to quickly duplicate them. Figure 7.2 illustrates a magnetic board and magnets used in the instruction.

Although several electronic devices that help the blind to carry out activities such as reading have been developed for decades, many students around the world still do not have access to this type of material.

## WHAT IS LIGHT? EXPLORING THE CONCEPT

A fact that must be considered when learning for a blind or visually impaired student is the time needed for the student to collect and process information. Visual acquisition of a concept is relatively quick when compared to other methods. Tactile or audible methods take time and have limitations. When learning about something new, the blind student must have time to explore all parts of the object.

After a succinct presentation on the subject, students are encouraged to learn through their own experience. Blind and visually impaired students need a succinct verbal description of everything. The teacher must refer to everything and everyone by name or description and not point out or use vague terms such as "this" or "that". During the presentation, speech must be clear to minimize any difficulties that students may have in following the reasoning.

Sighted students can also assist their disabled peers by explaining what is happening during exploration. It is highly recommended that blind students have a sighted partner in the laboratory.

Some activities are suggested by the teacher that will help students acquire new experiences for later extension activities. During this phase, students receive only minimal tutoring, and students are encouraged to explore new concepts on their own. During the exploration activity, the instructor provides encouragement, tutorials, and/or suggestions to maintain an appropriate level of mental imbalance. This activity provides information to the teacher regarding the students' ability to deal with the concepts and/or skills that are being introduced. Furthermore, students will deal with reasoning skills that can lead them to find a solution to a problem.

For example, to introduce the first concepts about light, in this first phase students are exposed to sunlight. The student feels the interaction of the sun's rays with their skin. Touch is one of the main forms of interaction between blind students and the world. Some questions may be raised by the teacher at this time: what do you feel? Now, the student is led into a shadow, and a new question is raised: Do you notice the difference? Alternatively, the teacher can use a laser of reasonable power to sensitize the skin and, through the heat generated, the student can feel the light. However, it is important to note that moderate and high-power lasers are potentially dangerous because they can burn the retina, or even the skin. To control the risk of injury, standards have been created, for example ANSI Z136 [12] in the United States and the international standard IEC 60825 [13]. These standards also describe necessary safety measures, such as labeling lasers with specific warnings, and wearing laser safety glasses when operating lasers.

For an inclusive teaching, given the need for more time on the part of the student with VI, one suggestion is to use alternative times so that this student can be attended to individually to be introduced to the equipment and methods that will be covered in class subsequent.

## WHAT IS LIGHT? DEVELOPING THE CONCEPT

The use of concrete examples is essential for blind students to learn. Solely verbal descriptions do not provide the same quality of information as when accompanied by tactile objects. To aid student understanding, concept development should be related to students' daily experiences. In the second phase, the concrete experience provided in the previous step is used as the basis for introducing a concept. The roles of the student and instructor in this activity may vary depending on the nature of the content. Generally, students should be asked to "invent" part or all the relationship for themselves. The instructor provides encouragement and guidance to the student when needed. This procedure allows students to self-regulate and therefore move toward balance with the concepts presented. During the model construction activity, students are encouraged to formulate relationships that generalize their concrete ideas and experiences. The teacher acts as a mediator helping students to formulate these relationships in a way that is coherent with the objectives of instruction.

For example, the teacher can help the student by introducing light as the intermediary agent in the distance interaction between a source (the sun or laser pointer) and the receiver (the student's skin). Here, the operational definition for light as "the radiation detected by the human eye" does not make sense to the blind student. Building on the previous step, we suggest a new definition: "Light is a form of radiant energy that impresses your skin by touch." The student feels the interaction of radiation with their skin. We know that touch is one of the main forms of interaction between blind students and the world. Touch is one of the five human senses, usually divided into four systems: thermoception (perception of temperature), somatosensory (identification of textures), proprioception or kinesthesia (recognition of the body's spatial location), and nociception (perception of pain).

Returning to the concept of light, none of us have direct access to reality. What we do is interpret the experiences we have with reality. Therefore, the operational definition for the light is as good as the standard definition. As Frish [16] states:

> ... We shouldn't ask what light really is. Particles and waves are both constructs of the human mind, made to help us talk about

> the behavior of light in different circumstances. With Bohr we abandon the naive concept of reality, the idea that the world is made of things, waiting for us to discover their nature. The world is made by us, by our experiences and the concepts we create to connect one to the other.

The rectilinear propagation of light can be easily explored with the help of a laser. On the bench, the teacher can place a Styrofoam. The student can identify the point at which the laser light sensitizes their skin. With the help of a pin, the student fixes the pin on the Styrofoam and by repeating this process several times he can check that the pins he fixed are aligned on the same straight line. The laws of reflection and refraction can be easily demonstrated.

## THE RAY MODELS: EXTENSION OF THE CONCEPT

In the third stage of the learning cycle, the student is allowed an opportunity to directly apply the concept or skill learned during the concept creation activity. This activity allows additional time for the necessary accommodation for students who need more time to achieve mental balance. It also provides additional information in the form of balancing experiences for students who have already accommodated the concepts presented. To begin the concept extension activity, students and teachers interact in planning an activity to apply the developed concept and/or skill, in a situation related to the instructional objectives. Although this activity allows students to extend the concept developed to apply it directly in a new situation, the nature of the expansion activity provides for a balance of new cognitive skills.

## ACCESSIBLE DRAWINGS AND DIAGRAMS

In science, technology, engineering, and mathematics (STEM) classes, particularly in geometrical optics, visual aids like charts and graphs are crucial. Reading, interpreting, and creating tables and graphs play a fundamental role in education. However, for visually impaired students, this mode of information is demanding. Their experience of the world differs from that of fully sighted individuals. Additionally, they face the obstacle of being unable to access printed information. According to the American Foundation for the Blind in 2017, in Western cultures, 75% of visually impaired children attend regular primary or secondary education, while the remaining children receive specialized education. In some countries, inclusive education laws mandate that schools admit children with special needs into regular educational programs and provide the necessary additional support for their success.

But how can we create drawings and diagrams accessible to visually impaired individuals? Certain criteria must be met: materials should be readily available, lines should be easily removable or erasable, and facilitate quick reproduction. Moreover, lines should be raised to enable easy differentiation for disabled students. Magnets and magnetic boards fulfill these criteria.

For instance, Figure 7.1 shows an $80 \times 50\,\text{cm}^2$ lightweight magnetic board and various types of magnets, predominantly rubber magnet strips (1.0 cm wide and 3 mm thick) used to teach optics to the visually impaired. These strips are highly practical for forming Cartesian axes and straight lines, such as those depicting the linear propagation of light (see Figure 7.1). Such dimensions facilitate tactile recognition for blind students. Other magnet types, like mats and shaped disks, with reasonable thickness, are utilized for drawing curved lines and marking points on graphs.

Unlike sighted students, who grasp a picture instantly, blind students 'read' the picture step by step. They use both hands to trace the raised lines on the board. Initially, students face challenges in interpreting the board's

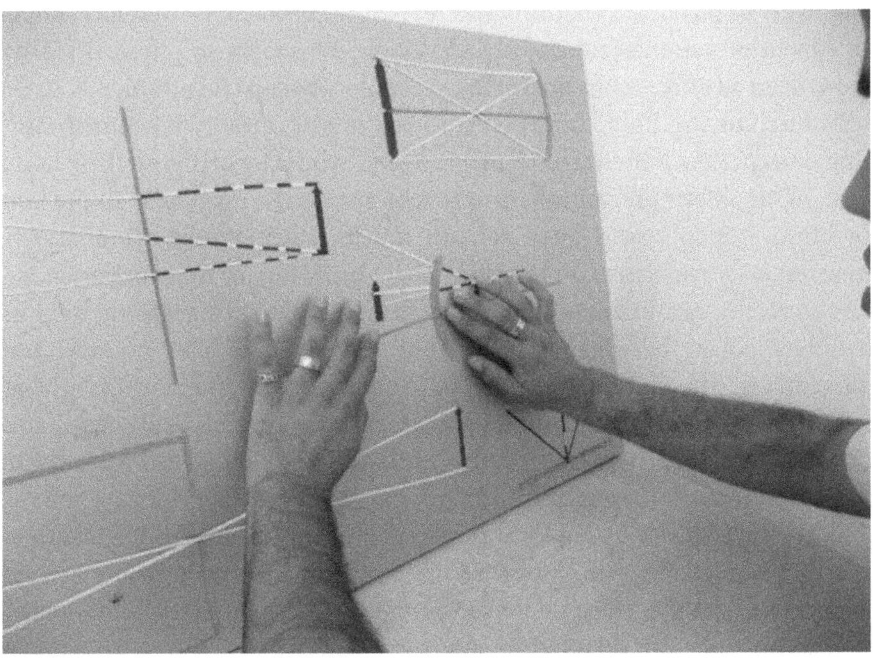

FIGURE 7.1 A visually impaired student learning optics by using a magnetic board and various types of magnets.

pictures. Interpreting pictures requires a learning curve where students associate picture elements with real objects.

The simplest depiction of light is as straight-line rays. For sighted individuals, rays represent the transmission of light through a medium with particles. But how do we convey this to blind individuals? One solution is through abstract descriptions, like mathematical lines, as depicted in Figure 7.1.

To depict light refraction, we easily draw broken and curved lines using magnetic strips. These drawings are stable enough for students to follow with their fingers. It's crucial to note that magnets and objects merely establish similarities, representing common properties of real objects. These "object-models" form conceptual images, abstractly representing elements within the real system for interpretation. This approach is grounded in the idea that teaching physical phenomena should be supported by creating conceptual models. Since modeling theory involves procedural knowledge, it's best absorbed through specific modeling activities, where theory gradually unfolds to oversee and direct these activities. Active learning, facilitated by model construction, requires syncing observed events with the scientific method, rather than just accumulating facts and formulas, structuring scientific theories based on modeling principles to apply them effectively to physical situations and real-world problems. Therefore, physics education emphasizing model development adapts to physical reality and is seen as effective in nurturing students' physical intuition.

In introducing light concepts, students might initially encounter sunlight or experience the warmth of a lamp or candle by bringing their hand close, with safety precautions to prevent accidents, especially important for blind students who often rely on touch. Additionally, a moderately powered laser can sensitize the skin. Then, building on these experiences, new concepts can be introduced. Teachers can assist by framing light as an intermediary between a source (like the sun or a laser pointer) and the receiver (the student's skin). The conventional definition of light as "radiation detected by the human eye" doesn't apply to blind students, so a new definition is proposed: "Light is radiant energy perceived by the skin through temperature sensation."

With careful attention to the safety precautions mentioned earlier, teaching the linear propagation of light can be effectively demonstrated using a laser device. The student can pinpoint where the laser light touches their skin, typically denoted by a specific sensation. Using a Styrofoam sheet on the bench, the teacher can guide the student to mark these points with pins (see Figure 7.2). Repeating this process multiple times allows

FIGURE 7.2 Scheme of a sheet (Styrofoam for instance). With the aid of the teacher, the student can mark light trajectories with pins.

the student to trace the path of light, confirming that the pins align in a straight line. Similarly, the laws of reflection and refraction can be explored by using pins as markers on a Styrofoam sheet. Throughout these activities, students are encouraged to derive general relationships from their experiences. The teacher serves as a facilitator, aiding students in articulating these relationships to align with the instructional objectives.

## DISCUSSIONS

Models serve as essential tools for science educators, given that modeling lies at the core of scientific inquiry. A physical model provides a simplified, objective representation of a phenomenon, aiding our comprehension of its underlying principles. For visually impaired individuals, modeling acts as a substitute for direct observation and experimentation, particularly evident in the traditional teaching of geometric optics, where lenses are utilized to study light refraction and reflection. Diagrammatic models become invaluable when creating experimental conditions is impractical due to physical limitations. However, it's crucial for instructors to acknowledge that achieving a completely accurate representation for students is often unattainable. Models, mockups, or diagrams function as simplified simulations of reality, designed to elucidate the phenomenon they depict. Nonetheless, analog models inherently possess limitations; for instance, light is sometimes conceptualized as a wave and other times as a particle.

Contemporary educational methodologies advocate for student-centered teaching, acknowledging the importance of students' prior knowledge in the learning process. We have adapted learning circles, drawing from Piaget's theory of intellectual development, which encompass three key phases: exploration, concept introduction, and concept application.

During the exploration phase, students engage with unfamiliar situations, generating questions that challenge their existing patterns of reasoning. For example, students may be prompted to approach a candle flame or sunlight to sense the light through their skin, prompting inquiries such as 'What is light?' or 'How does it propagate?' Teacher intervention is minimal during this phase.

The concept of light is then introduced in the second phase, requiring more teacher interaction. However, conventional definitions of light, tailored to sighted individuals, become irrelevant for blind students. Instead, a new representation of light as a sensation on the skin is necessary. Students may explore tactile representations like a magnetic board to grasp these concepts.

In the concept application phase, students apply their newfound understanding to various situations, emphasizing physical experience and social interactions with peers and teachers.

The positive outcomes of this approach indicate that students can comprehend and engage with presented figures and graphics in detail, fostering their motivation to continue learning and ensuring a successful teaching-learning process.

In an inclusive class, the use of these tools (board with magnets and laser) can bring extra benefits to other students, as they are playful tools and therefore arouse curiosity and interest in manipulating them. As well as the learning cycle methodology favors inclusive teaching, benefiting all students in the class.

## CONCLUSIONS

The discussion of the learning of concepts by visually impaired people fundamentally permeates everything that alludes to the learning of concepts by any person. Several studies indicate cognitive abilities of visually impaired people that are like those of sighted people. The differences are associated with alternative modes of cognitive processing of sensory information. It is a known fact that one of the greatest difficulties encountered by a visually impaired physics student is related to carrying out experiments and, in the case of optics, their mental representation of what light might be. Therefore, in this work, emphasis was placed on the use of magnetic boards and magnets by students with visual impairments, so that their learning in mathematics and physics becomes more beneficial. During the instructions, it was observed that the visually impaired student establishes his relationship with physics learning through a continuous process of personal overcoming. Learning cycles provide an opportunity to argue and test alternative concepts, becoming unbalanced—in the Piagetian sense—acquiring more appropriate concepts as well as more elaborate reasoning patterns.

It is fundamental to prepare physics educators to be adaptable and innovative in their teaching approaches, especially when addressing the needs of visually impaired students. By moving away from a sole reliance on visual methods and embracing diverse instructional strategies, teachers can create more inclusive and effective learning experiences in the field of optics.

## BIBLIOGRAPHY

1. Costa, L. G., Neves, M. C. D., & Barone, D. A. C. (2006). O ensino de Física para deficientes visuais a partir de uma perspectiva fenomenológica, *Ciência & Educação (Bauru)*, *12*, 143–153.
2. Baughman, J. Jr., & Zollman, D. (1977). Physics labs for the blind. *The Physics Teacher, 15*, 339–342.
3. Cetera, M. M. (1993). Laboratory adaptations for visually impaired students: thirty years in review. *Journal of College Science Teaching*, 12, 384–393.
4. Dion, M., Hoffman, K., & Matter, A. (2003). A Teacher's Manual for Adapting Science Experiments for Blind and Visually Impaired Students, Texas School for the Blind and Visually Impaired. www.tsbvi.edu/Education/Manual2.doc
5. Eicherberger, F. J. (1974). Teaching Science to Blind Students. *The Science Teacher, 41*, 53–54.
6. Fletcher, R. C. (ed.) (1973). *The Teaching of Science and Mathematics to the Blind*. Oldbury: Royal National Institute for the Blind.
7. Weems, B. (1977). A physical science course for the visually impaired. *The Physics Teacher, 34*, 333–338.
8. Henderson, D. R. (1965). Laboratory methods in physics for the blind. MSc Thesis. University of Pittsburg.
9. Camargo, E. P., & Nardi, R. (2007). Planejamento de atividades de ensino de Física para alunos com deficiência visual: dificuldades e alternativas. *REEC*, 6, 378–401.
10. Azevedo, A. C. (2010), Construção gráfica para alunos cegos. *Física na Escola*, *11*, 48.
11. Parry, M., Brazier, M., & Fischbach, E. (1997). Teaching college physics to a blind student. *The Physics Teacher, 35*, 470–474.
12. ANSI 136.5 (American National Standards Institute) (2007). *American National Standard for the Safe Use of Lasers*. Orlando, FL: Laser Institute of America.
13. IEC 60825 (2007). *Safety of Laser Products Package*. International Electrotechnical Comission.

14. Camargo, E. P. D., & Nardi, R. (2007). Dificuldades e alternativas encontradas por licenciandos para o planejamento de atividades de ensino de óptica para alunos com deficiência visual. R*evista Brasileira de Ensino de Física, 29*, 115–126.
15. Camargo, E. P. D., & Nardi, R. (2008). O emprego de linguagens acessíveis para alunos com deficiência visual em aulas de óptica. *Revista Brasileira de Educação Especial, 14*, 405–426.
16. Frish, O. R. (1972). *The Nature of Matter.* London: Thames and Hudson.

CHAPTER 8

# Teaching modern physics

## INTRODUCTION

Colleges, universities, and high schools should provide equitable facilities that enable students with disabilities to access equal educational opportunities. Despite the central importance of physics as a foundational subject in STEM education, there is a notable lack of resources for teaching physics to students who are blind or visually impaired. In this chapter, we share insights gained from developing an accessible modern physics curriculum for a blind student.

The focus is on actionable steps instructors can take to ensure accessibility in areas such as class sessions, course materials, tutorials, and demonstrations, rather than administrative accommodations like extended testing time.

In the ideal scenario, when an institution is informed that a blind student will be attending physics classes, the course instructor and the disabilities coordinators, if the institution has them, should meet well ahead of the classes start to discuss course logistics. This early planning allows adequate time to establish an effective instructional and support team, ensure that essential assistive technologies are available, make class sessions accessible, and prepare course materials in accessible formats [1].

DOI: 10.1201/9781003510079-8

The key decision is whether to adopt individualized, also known as segregated, integrated, or inclusive instruction. In individualized (segregated) instruction, the student and instructor engage in one-on-one tutorials [2]. Segregation is the practice of separating students with different backgrounds or disabilities into separate environments. Segregation occurs when special needs schools are the primary way to educate people with disabilities. Integration means that policies are used to address the needs of people with different backgrounds, but these groups are often absorbed into the mainstream and treated as the "diverse" group. Inclusion means a welcoming and equal treatment of individuals and groups with different backgrounds. Inclusion aims to benefit all involved and ensure clear participation. In an inclusive (mainstream) approach, the student participates in regular class sessions alongside peers. Instructors should collaborate with the student to decide on the most appropriate method, considering the students' specific needs and the resources available within the institution.

Brockington and Pietrocola [3] point to the need to include modern and contemporary physics in high school. The work resulting from almost two decades of educational research is sufficient to ensure the need to update physics programs in high school. However, there is little research related to the implementation of topics of modern and contemporary theories. In the case of teaching modern physics to the blind it is even worse, there is a huge lack of studies devoted to it. Brockington and Pietrocola emphasize that the challenges are imposed not only by the intrinsic complexity of these topics, but also by an insecurity inherent in any attempt to change the school domain. Most teachers are stuck in a pedagogical environment without much flexibility, whether due to content prescriptions, restricted schedules and specificities of their own disciplines.

Pietrocola [4] identified four types of didactic-epistemological obstacles associated with modern physics: phenomenology, language/formalization, conceptual structure, and ontological base. These obstacles stem from the challenges observed in developing educational content for high school instruction. (I) *Phenomenology*: In classical physics, many phenomena can be observed in everyday life or demonstrated in simple experiments within didactic laboratories. However, the phenomena studied in modern and contemporary physics—such as those involving the very small or very fast—exist beyond the scope of daily experience. These phenomena are neither accessible in everyday contexts nor easily replicated through simple experiments. For example, there are no readily available tools to explore the dual nature of light in a comparable manner. (II) *Language/*

*Formalization*: The contents of classical physics can be adapted for high school teaching through simplified mathematical formalism, relying on basic algebra and geometry. In contrast, modern and contemporary physics require advanced mathematical tools, which are not easily integrated into high school curricula. Two approaches have been explored in the literature: requiring students to develop the necessary technical expertise or prioritizing conceptual and qualitative physics through metaphorical or analogous explanations. (III) *Conceptual Structure*: Classical physics concepts often extend from intuitive, everyday ideas—examples include force, temperature, heat, and energy. These concepts have been central to research on misconceptions and cognitive development. Modern physics concepts, however, frequently defy intuition and common knowledge. Terms such as probabilistic determinism, orbital positions, spin, reduced mass, and relative space and time may seem relatable to intuitive ideas but ultimately diverge significantly, creating what some have called "old language graveyards." (IV) *Ontological Base*: Classical physics entities are rooted in perceivable objects from everyday life, such as particles, waves, space, and time. In contrast, modern and contemporary physics deal with entities like massless particles, quantum energy, virtual particles, and curved space. These concepts exhibit unique properties and behaviors that are far removed from the familiar objects of daily experience.

*Didactic-Pedagogical Obstacles:* Pietrocola [4] defines *didactic-pedagogical obstacles* as systemic conditions in the educational process that hinder the introduction of modern and contemporary physics content. While they have facilitated the establishment of classical physics in classrooms, they also create significant barriers to incorporating modern physics into the curriculum.

The inclusion of modern physics in the curriculum presents a challenge in designing appropriate teaching scenarios. These topics involve phenomena and entities that are not directly perceptible in everyday life and can only be understood through highly rational and technical methods. Thus, the main challenge in teaching this subject lies in creating conditions that enable both teachers and students (visually impaired or not) to overcome the difficulties associated with perceiving and understanding these abstract concepts [5].

Teaching modern physics to blind students presents extra unique challenges due to the visual nature of many concepts and experiments in the subject. However, with appropriate strategies and tools, these difficulties can be mitigated. (I) *Visual Dependence of Concepts:* Many physics concepts, such as graphs, diagrams, waveforms, rely heavily on visual representations. In addition, visualizations of abstract concepts like fields, vectors, or particle behavior can be hard to convey non-visually. (II) *Experimental Limitations:* Traditional modern physics experiments often rely on visual cues, such as reading measurements from instruments or observing phenomena. (III) *Mathematical Representations:* Physics involves equations, graphs, and spatial reasoning, which can be difficult to interpret using non-visual methods. (IV) *Lack of Accessible Materials:* Many textbooks, instructional materials, and online resources are not adapted for blind students, particularly for advanced topics. (V) *Teacher Preparedness:* Many educators lack training in how to adapt physics content and activities for blind students. (VI) *Classroom Integration:* Blind students may feel excluded from group activities or discussions if accommodation is not provided.

Physics education research has identified some clues to success in teaching modern physics at the introductory level: (i) focus the presentation on phenomena and practical applications, (ii) maintain a manageable level of mathematical complexity, and (iii) provide sufficient time for understanding. These are novel concepts that require careful exploration and cannot be hurried.

The aim of this chapter is to familiarize students, for whom this is their final physics course, with fundamental phenomena, concepts, and terminology on modern physics.

## BUILDING MODELS ABOUT THE ATOM

While all students are familiar with terms like "atom" and "electron," most have only a vague understanding of their meanings. They often carry a mix of partially recalled "facts" from chemistry but lack a clear mental image of what an atom truly is.

The importance of teaching about atoms is directly linked to the basic understanding of science and the development of critical skills for scientific thinking. Atoms are the building bricks for matter. Knowledge about them is essential to understanding natural phenomena, technological advances, and chemical and biological processes that shape our world. The concept of atoms is fundamental to physics, chemistry, and biology. Without an understanding of their structure and behavior, it is impossible to understand the laws of nature, such as chemical interactions, conservation of mass, radioactivity, and even biological processes, such as the functioning of DNA. Teaching atoms promotes logical and scientific reasoning skills. Students learn to think on a microscopic scale, extrapolate visible phenomena to the invisible, and accept evidence-based explanations, even if these are not intuitively obvious. Many technological advances derive from atomic knowledge, such as semiconductors, lasers, nuclear power, and nanotechnology. By learning about atoms, students understand the principles that underpin these technologies and their impact on everyday life. Atoms are at the center of debates about energy and the environment. Nuclear power, for example, relies on the fission of atoms, while atmospheric chemistry explains the impact of polluting gases at the molecular level. Teaching about atoms helps to develop citizens who are capable of understanding and participating in informed discussions about sustainability. Understanding that everything in the universe is made of atoms—from the air we breathe to the stars—sparks students' natural curiosity. This fascination with the invisible world can inspire future generations to explore careers in science and technology. The study of atoms is crucial to fields such as medicine (magnetic resonance imaging, radiation therapy), engineering (advanced materials), agriculture (fertilizers, pest control), and many others. It connects disciplines and enables an integrated understanding of the world. Teaching about atoms also provides an opportunity to explore the history of science, from Greek philosophy to the discoveries of Dalton, Rutherford, Bohr, and others. This highlights how human knowledge evolves over time based on observation, experimentation, and innovation. Teaching about atoms is essential to developing individuals who understand the world at a fundamental level. Such education promotes critical thinking, fosters curiosity, and prepares people to deal with scientific and technological challenges in an informed and creative way. It also helps build a more conscious and future-ready society.

This section reviews the literature [6–9] and focuses on presenting and discussing tactile materials designed for teaching atomic models and electron distribution to blind students. These materials, developed for use in basic education, promote inclusivity by allowing both blind or visually impaired students and sighted students to engage with them, fostering learning. Tactile resources can play a significant role in helping blind students understand atomic structure. Additionally, their affordability makes them easily reproducible and accessible for chemistry teachers in basic education.

When teaching about atomic structure, one can start by discussing Greek atom. Then, one can move to John Dalton's model (1766–1844) and the Crookes tube, which later led to Joseph Thomson's atomic model

(1856–1940). Following this, the gold foil experiment, which contributed to Ernest Rutherford's model (1871–1937), and the introduction of quantum chemistry with Niels Bohr's quantized atom (1885–1962) can be explored.

Next, the instructor can introduce electronic distribution using a tactile and concrete representation of Linus Pauling's diagram (1901–1994). It's essential for blind learners to validate the developed materials, ensuring they can be refined for future use and broader dissemination in basic education.

Velloso et al. [9] presented a proposal for approaching the contents of physics to a blind student who presented many conceptual difficulties during the COVID-19 pandemics. Due to the need for social isolation, the lectures had to be remote, and all materials produced and delivered to the student's home were needed to take this new reality into account. The authors ensured that the course was as similar as possible to the in-person

FIGURE 8.1 Top left: a tactile-visual representation of a negative charge, with a gray ball further away from the center of the configuration and on the right side a positive charge, represented by the hole left by the lack of a ball. Top right: The student manipulates the two toy atomic models. Bottom right: Tactile-visual representation of the hydrogen atom. Bottom right: Tactile-visual representation of an arrangement of atoms.

course. They prepared several adapted materials and delivered them to the student's home after being sanitized with 70% alcohol. One kit of materials sent to the student contained models with tactile-visual representations made of Styrofoam and wooden popsicle sticks. The models represent the hydrogen atom (Figure 8.1), atoms with excess electrons, and atoms with a lack of electrons, or "holes." They also created a structure that represents an arrangement of atoms (bottom part of Figure 8.1), containing two small balls that represent the "free" electrons that had a very weak bond with the nucleus and that could detach from the attraction of this atom and join the neighboring atom. To represent this weak bond, they inserted small magnets inside the Styrofoam balls so that, when bringing them close to each other, the student could see that they were held together by a very small force. The instructor carefully explained why the balls attracted each other. Thus, by analogy, the smaller balls represent the free electrons and the larger ones the atoms. These smaller balls were attached to nylon threads so that they would not fall to the floor during the student's activity.

## TEACHING RELATIVITY

The importance of teaching relativity, both special and general, goes beyond simply understanding advanced physics concepts; it lies in its transformative impact on understanding the universe and developing critical and innovative thinking. The Theory of Relativity, proposed by Albert Einstein, revolutionized our understanding of space, time, gravity, and energy. Teaching these concepts helps students break away from classical and intuitive views, such as the idea that time is absolute or that space is a fixed stage where events occur. This new perspective stimulates curiosity and fascination with the cosmos. Relativity challenges concepts that seem "obvious" and require students to think abstractly and logically. Understanding how time can dilate or how gravity is the curvature of space-time requires rigorous analysis and an open mind to new ways of thinking. Relativity has direct implications for everyday technologies, such as GPS, which need to correct for the effects of time dilation caused by the speed of satellites and Earth's gravity. Teaching relativity shows students how theoretical science can translate into practical technological advances. In a world where science and technology are constantly evolving, understanding relativity prepares students to tackle broader issues, such as space travel, black holes, nuclear energy, and even the exploration of other planets. Relativity is a fascinating topic that can attract young people to the fields of physics, astronomy, engineering, and mathematics.

It challenges the imagination and inspires future generations to explore the limits of human knowledge. In addition to scientific concepts, teaching relativity also allows us to reflect on how science advances: through questioning established paradigms, collaborative work, and the tireless search for truth. Einstein is an example of how genius and persistence can change the world. Teaching the Theory of Relativity is not just about teaching physics; it is an opportunity to educate citizens with critical thinking, capable of innovating and understanding the depth and beauty of the universe. It is investing in the intellectual and cultural development of future generations.

Granhen and Sobral [10] report on the teaching of the concepts of time dilation and spatial contraction, present in Einstein's Special Relativity, to a sixteen-year-old student with low vision. To teach the concepts of time dilation and space contraction, two experiments were proposed, the first called the "time dilation screen" and the second the "space contraction network." The aim was to address concepts of special relativity in a playful way, specifically the topics of time dilation and space contraction. The first device, the screen, consists of an adapted table containing a screen with the figure of a spaceship drawn on a canvas that moves when an on-off switch is activated. And the second device, the net, is built on a metal skeleton in the shape of a table without the top, with crossed elastic strips fixed on top, forming identical squares like a net, with plastic bottle caps at the crossing points.

## TEACHING COMPTON EFFECT

When teaching Compton and photoelectric effects (see next section), one should have some learning objectives: The goal is to identify phenomena that classical physics cannot explain, highlighting the necessity for a new theory. It also aims to provide experimental evidence confirming the existence of atoms and their properties. Additionally, it seeks to comprehend Compton and photoelectric effects and their significance, understand the photon model and its uses, and grasp the evidence for matter waves and the de Broglie wavelength.

Electrons, protons, and other subatomic particles, as well as the particle-like behavior of light, are not directly perceivable by our senses. However, they can be made accessible through technical means. The advanced equipment used in modern science, particularly in the realms of atomic and subatomic phenomena, enables us to indirectly observe phenomena that are both rationally predicted and technically constructed.

Various types of epistemological shortcomings are inherent in the process of didactic transposition, particularly concerning the organization and evolution of scientific knowledge itself [5]. Suggesting the presence of didactic-epistemological challenges specifically linked to the didactic transposition of modern physics, as distinct from those associated with classical physics, leads to the following hypothesis:

Rodrigues and Ferreira [11] suggested the use of three-dimensional (3D) model to demonstrate the Compton effect to students with visual impairments. Figure 8.2 illustrates the Compton effect. It takes place when an x-ray photon hits a bound electron. Consequently, the ionization of the target atom takes place, a change in the x-ray direction, and a reduction in the energy of the photon. The wavelength of the scattered photon is larger than that of the incident photon. But how should we draw the diagrams so that the visually impaired student can feel them tactilely? Some criteria must be followed: the material must be easily accessible, the lines, diagrams, and drawings must be easily erasable or removed and allow for their rapid duplication by the student.

Rodrigues and Ferreira [11] used low-cost materials to construct the model, such as a transparent acrylic board where the model is built, clay, colored fabric paints to represent the wave character of the photon, and

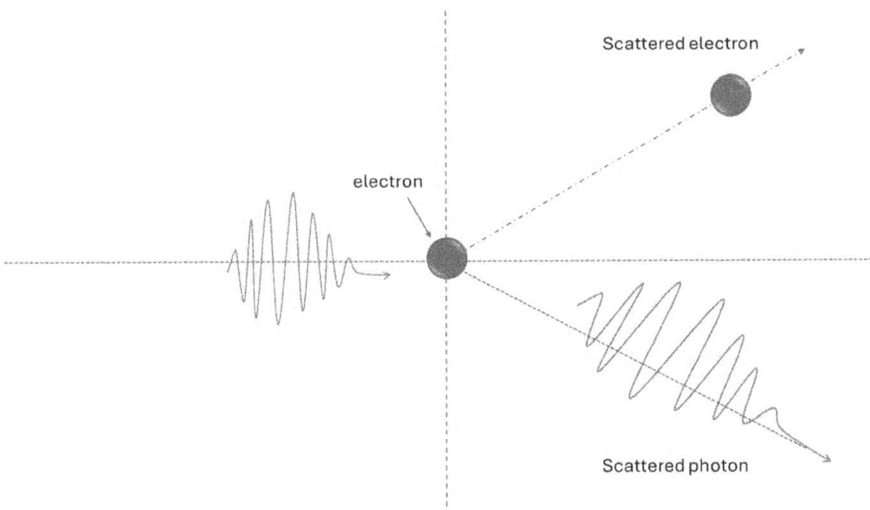


FIGURE 8.2 The Compton effect takes place when an x-ray photon hits a bound electron. It causes the ionization of the target atom, a change in the x-ray direction, and a reduction in the energy of the photon. The wavelength of the scattered photon is larger than that of the incident photon.

glue. The project was implemented in a class for high school students with visual impairments. During the class, the experiment on the Compton effect was used, and a questionnaire in braille was administered to verify the students' understanding of the content taught.

## PHOTOELECTRIC AND PHOTOVOLTAIC EFFECTS

### Photoelectric effect

According to some authors [12], presenting the story of black-body radiation in an introductory course in an intellectually honest and meaningful manner is impractical, as oversimplification often leaves students confused. In compliance with it, the most straightforward and comprehensible approach to introducing the quantum concept is through the photoelectric effect, a method nearly universally adopted in modern textbooks. A true understanding of modern physics requires students to grasp how and why the existence of atoms is established. Facts and definitions committed to memory without a connection to physical phenomena do not constitute genuine understanding.

The reproduction and use of diagrams and symbols in tactile formats, such as high-relief constructions, is of paramount importance when teaching modern physics to the visually impaired student. These tactile tools enable visually impaired students to perceive objects often used to illustrate problem scenarios in modern physics through touch. Thus, one should introduce students to some of the experimental evidence that reveals the existence of atoms and photons. One also must help them understand that many phenomena related to light and atoms cannot be anticipated or explained through classical physics.

Teaching modern physics relies heavily on visuals to explain the concepts. Since understanding the figures and texts is essential for effective learning, it is crucial to address the gap where only text is translated into Braille, while figures are left for teachers or aides to interpret for visually impaired students.

Given that images play a significant role in conveying meaning and facilitating understanding in scientific materials, alternative resources must be provided for those unable to engage with visual content. Images are a widely accepted medium in scientific communication, offering unique insights into nature and helping to illustrate abstract ideas.

If textbooks adapted for visually impaired students omit most of their images, these materials lose much of their utility, especially in fostering the independence that is critical for student development. To support the

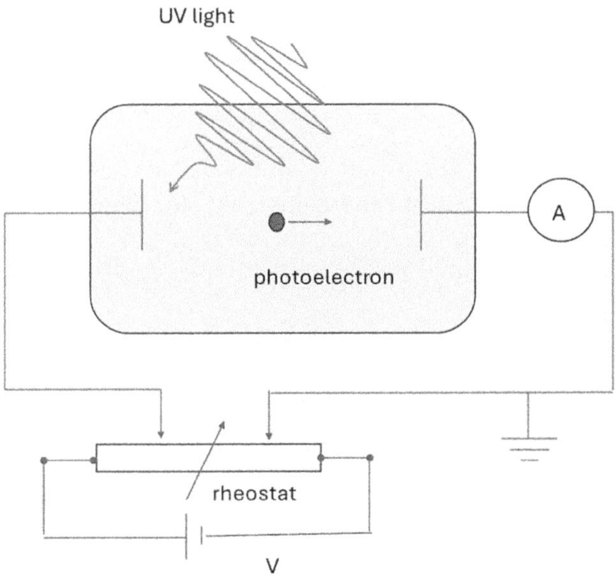


FIGURE 8.3 Schematic diagram of the photoeffect apparatus.

autonomy of visually impaired students, it is essential to improve their access to the full range of content in textbooks and other educational materials included in the curriculum. Figure 8.3 shows a schematic diagram of Lenard's apparatus for measuring the photoelectric effect. This diagram can be easily reproduced in a tactile format by the teacher.

In 1905, Einstein suggested an explanation for the dependence of photoelectric emissions on the frequency of radiation. He suggested that free electrons interact with electromagnetic radiation in the same manner that Planck described for atomic oscillators in the context of black-body radiation. According to Einstein, the energy E absorbed by an electron from electromagnetic radiation of frequency $\nu$ is given by $E = h\nu$, where h is Planck's constant.

In metals, some electrons are relatively free to move within the material. However, at normal temperatures, these electrons cannot escape the metal because they lack sufficient energy to overcome the Coulomb potential barrier at the surface. The minimum energy required for an electron to escape from a specific metal is called the work function, denoted as W. When an electron absorbs energy $E = h\nu$, the excess energy, after overcoming the work function W, manifests as the kinetic energy K of the escaping electron. The energy required for electrons to escape from the metal is not the same for all electrons. Since the minimum energy to escape is W, the maximum kinetic energy of the electron can be expressed as follows:

$$K_{max} = hn - W \tag{8.1}$$

From Eq. (8.1), we see that, at the frequency $\nu_o$, for which

$$hn_o = W \tag{8.2}$$

The maximum kinetic energy of the electrons becomes zero at a specific frequency. This frequency, $\nu_0$, is the threshold frequency—the minimum frequency required for photoelectric emission to occur. For frequencies below $\nu_0$, no emission takes place because the electrons cannot absorb enough energy in a single interaction to escape from the metal, regardless of the radiation's intensity.

Einstein's explanation effectively accounts for the observed dependence of the photoelectric effect on the frequency of radiation. The maximum kinetic energy, $K_{MAX}$, can be measured using the setup shown in Figures 8.3 and 8.4. By applying a potential difference V between the plates, the motion of the photoelectrons can be slowed. At a specific voltage, $V_0$, the electrometer (A) registers a sudden drop to zero current, indicating that even the fastest electrons are no longer reaching the plate. It should be stressed that work function is energy, usually measured in electron volts (eV). The student can

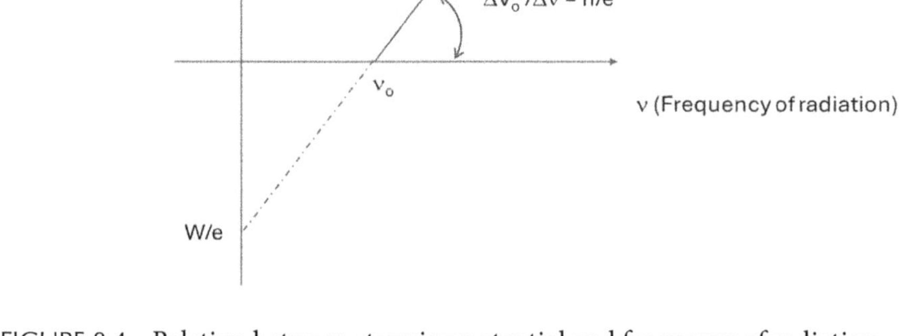


FIGURE 8.4 Relation between stopping potential and frequency of radiation.

get confused by the difference between volts (V), which measures electrical potential, and electron volts which is a measure of energy.

Classroom activities should focus on reasoning through both the experiment and the theory. Research suggests that exploring the conceptual aspects of the physics behind the effect before tackling numerical problems is more effective for teaching.

The presentation of the photoelectric effect should place greater emphasis on explaining the experiment than is typically done. The instructor should clearly outline the findings from Lenard's experiment and the theory developed by Einstein. Furthermore, the teacher should provide a clear explanation of the photon model and how it is applied to the experiment.

### Photovoltaic effect

Renewable energy is a highly relevant topic today, particularly in the context of global warming. Given its significance, it is essential to address this subject in schools within an inclusive education framework. Hands-on experimental activities are important to foster effective conceptual discussions among students, but enhancing the teaching-learning process is crucial to align with inclusive education principles. Hahn et al. [13] present a low-cost experimental activity using Arduino, designed to enable both regular and blind students to collaboratively explore and understand key aspects of solar energy and photovoltaic panels. The activity demonstrates concepts such as how the electrical power generated by a photocell varies with the angle of solar light incidence and the load resistance in the circuit.

Usually, students are guided to analyze how the system's independent variables influence the energy output of a photovoltaic cell (Figure 8.5). To accomplish this, they typically assemble the experimental setup, using components such as a photovoltaic cell, voltmeter (V), ammeter (A), switch, connecting wires, and an incandescent lamp as a light source to simulate natural sunlight as closely as possible. The power generated by the system is calculated mathematically as the product of the voltage measured by the voltmeter and the current measured by the ammeter.

When the electrical circuit is closed and the light source is activated, the photovoltaic cell converts light energy into electrical energy. The amount of energy generated depends on the angle at which the light rays strike the photocell's surface and the load in the external circuit.

To design an activity that includes blind students in discussions about electric energy generation from solar power, the following objectives were established by the authors [13] (A) enable all students, including those with

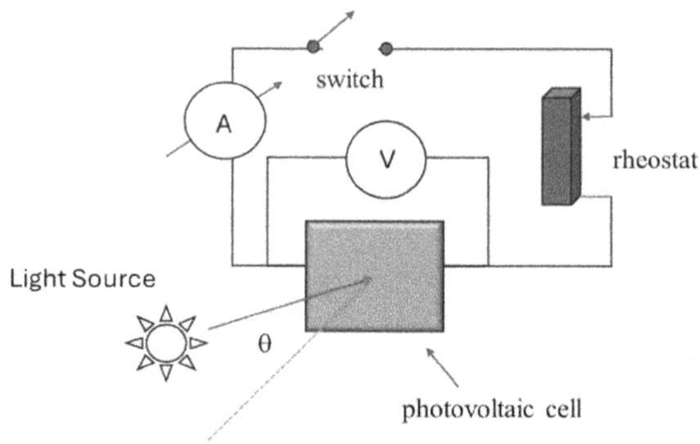


FIGURE 8.5 Schematics of an assembled electrical circuit, where V stands for a voltmeter and A for an ammeter.

**Source: The authors.**

visual impairments, to clearly perceive the effect of the angle of incident light on the solar panel; (B) ensure that blind students can engage in a process analogous to graphical construction, using objects or signals that provide a comparable understanding of the physics principles involved in the experiment; and (C) provide experimental data in a digital format accessible to all students through digital text-reading software.

The aim is to equip blind students—who naturally rely more on other senses such as touch, hearing, smell, and taste for perceiving the world with tools that leverage their heightened sensory abilities. The proposed adaptations to the experimental setup build on these strengths, offering both qualitative and quantitative learning opportunities.

The objective is to provide tools for blind students who, out of necessity, rely on their other senses (haptic perception, smell, taste, and hearing) to perceive the surrounding world with greater intensity than individuals without visual impairments. Consequently, the proposed adaptation in the experimental setup leverages this enhanced sensory perception, resulting in both qualitative and quantitative outcomes: To qualitatively measure the electric power generated by the panel, an Arduino microprocessor is used to convert the electrical signal into sound, where the pitch corresponds to the signal's strength—the higher the pitch, the greater the power produced by the photovoltaic panel. The angle of incidence of light can be measured quantitatively by extending the protractor and labeling the corresponding angles in Braille, enabling blind students to read and interpret the values.

The authors enhanced the experimental setup shown in Figure 8.3 by integrating an Arduino microcontroller board into the circuit. This setup measures the voltage and current generated by the photovoltaic panel, displaying the values in a separate window within the Arduino. Additionally, a piezoelectric buzzer is included to emit sound frequencies corresponding to the power output of the photocell. The buzzer produces lower frequency sounds for low power output and higher frequencies for greater power, enabling blind students to perceive how the photovoltaic cell's power varies with different parameters. Furthermore, a screen reader program can be utilized, allowing blind students not only to hear the frequency changes but also to access the exact measured values.

In conclusion, achieving truly inclusive education—beyond mere integration—requires a fresh approach to experimental setups that accommodate all students, fostering their active participation. This inclusivity paves the way for a more equitable society in the future. Additionally, diversity within a classroom should not be viewed as a challenge to overcome but as an opportunity for educators to explore innovative methods of presenting content, enhancing their skills, and becoming even more effective professionals.

## COMMENTS

There are strategies and solutions to overcome the difficulties associated with teaching modern physics to the visually impaired. The instructor can rely on tactile and auditory representations, i.e., use tactile models, such as 3D-printed objects, to represent physical phenomena (e.g., atomic structures, wave patterns, or force diagrams). One should convert visual data into auditory formats, such as using sound to represent the frequency of waves or accelerations. The instructor can also make use of accessible technology, such as leverage screen readers, braille displays, and accessible software to interact with written materials and data; use specialized tools like talking calculators, accessible data loggers, and sonification devices for experiments.

The instructor may also adapt experiments, by modifying them to focus on tactile, auditory, or other sensory outputs. For example, use audible indicators for motion experiments, and allow students to feel the vibrations of sound waves or the heat transfer in thermal experiments.

Another challenge is mathematical accessibility. One strategy is to use braille notations for mathematical equations and provide step-by-step verbal explanations. Another option is to utilize software like MathPlayer [14] or LaTeX-to-Braille [15] converters to render mathematical content accessibly.

It is of paramount importance to train physics teachers to develop inclusive teaching strategies and collaborate with specialists in visual impairments. The instructor should also involve students in adapting materials to ensure their needs are met.

Finally, one should encourage group activities that include tactile and verbal communication to foster collaboration and understanding. The teacher should design assessments that accommodate non-visual methods, such as oral exams or tactile problem-solving exercises. By embracing these strategies, educators can make physics more accessible and engaging for blind students, ensuring that they have equitable opportunities to learn and succeed.

## BIBLIOGRAPHY

1. Holt, M., Gillen, D., Nandlall, S. D., Setter, K., Thorman, P., Kane, S. A., … & Supalo, C. (2019). Making physics courses accessible for blind students: strategies for course administration, class meetings, and course materials. *The Physics Teacher, 57*(2), 94–98.
2. Fernandes, J. M., Franco-Patrocínio, S., Zambelli, M. H., & de Freitas-Reis, I. (2017). A elaboração de materiais para o ensino de modelos atômicos e distribuição eletrônica para discente cego: produtos de um projeto PROBIC-JR. *Experiências em Ensino de Ciências, 12*(6), 95–108.
3. Brockington, G., & Pietrocola, M. (2005). Serão as regras da transposição didática aplicáveis aos conceitos de Física Moderna?. Investigações em *Ensino de Ciências, 10*(3), 387–404.
4. Pietrocola, M. (2017). Curricular innovation and didactic-pedagogical risk management: teaching modern and contemporary physics in high schools. *Crossing the Border of the Traditional Science Curriculum*. Brill. Rotterdam: SensePublishers.
5. Pessanha, M., & Pietrocola, M. (2017). Particle Accelerators and Didactic Obstacles: A Teaching and Learning Experience in São Paulo and Cataluña. *Crossing the Border of the Traditional Science Curriculum: Innovative Teaching and Learning in Basic Science Education*, 45–59.
6. Laconsay, C. J., Wedler, H. B., & Tantillo, D. J. (2020). Visualization without vision–how blind and visually impaired students and researchers engage with molecular structures. *Journal of Science Education for Students with Disabilities, 23*(1), 1–21.
7. Wedler, H. B., Boyes, L., Davis, R. L., Flynn, D., Franz, A., Hamann, C. S., … & Wang, S. C. (2014). Nobody can see atoms: science camps highlighting approaches for making chemistry accessible to blind and visually impaired students. *Journal of Chemical Education, 91*(2), 188–194.
8. Freitas-Reis, I., Fernandes, J. M., Franco-Patrocínio, S., Faria, F. L., & Carvalho, V. (2017). Adaptações táteis de modelos atômicos para um ensino de química acessível a cegos. *Enseñanza de las ciencias*, (Extra), 4015–4020.

9. Velloso, M., Arana, M., Acioly, V., & Santos, A. C. F. (2021). Abordagem da Lei de Ohm para Estudantes com Deficiência Visual no Ensino Superior. Latin-American *Journal of Physics Education, 15*(1), 10.
10. Granhen, E. R., & Sobral, E. da C. (2019). Trabalhando os conceitos de dilatação temporal e contração espacial para alunos de baixa visão. *Scientia Plena, 15*(7), 074808.
11. Rodrigues, J. P. V. W., & Ferreira, F. C. L. (2019). EXPERIMENTO DE FÍSICA MODERNA EM TRÊS DIMENSÕES (3D) SOBRE O EFEITO COMPTON. Seminário e Projetos de ensino (ISSN: 2674-8134), *3*(1). recuperado de. https://periodicos.unifesspa.edu.br/index.php/spe/article/view/567
12. Arons, A. B. (1996). *Teaching Introductory Physics.* New York: Wiley, 816 (1990).
13. Hahn, M., Cruz, F., Carvalho, P. S., Ferreira, V., Marques, F., & Macieira, C. (2021). Adapting a solar photovoltaic panel experiment for blind students. *Physics Education, 56*(3), 035023.
14. MathPlayer 4.0. https://mathplayer.software.informer.com/4.0/. Accessed on January/1/2025.
15. StackExchange. https://tex.stackexchange.com/questions/32609/latex-to-braille. Accessed on January/1/2025.a

# Index

For Product Safety Concerns and Information please contact our EU representative GPSR@taylorandfrancis.com
Taylor & Francis Verlag GmbH, Kaufingerstraße 24, 80331 München, Germany

www.ingramcontent.com/pod-product-compliance
Ingram Content Group UK Ltd.
Pitfield, Milton Keynes, MK11 3LW, UK
UKHW022123190926
475702UK00013B/1607